# SALT AS A FACTOR IN THE CONFEDERACY

*Ella Lonn*

# SALT AS A FACTOR IN THE CONFEDERACY

SOUTHERN HISTORICAL PUBLICATIONS #4

UNIVERSITY OF ALABAMA PRESS

Library of Congress Catalog Card Number: 65-19663
Manufactured in the United States of America by
Murray Printing Company, Forge Village, Mass.

0-8173-1269-2 (pbk: alk. paper)

# PREFACE

It is only when a prime necessity thrusts itself upon public attention by its absence that a person ceases to take it for granted. Only when he no longer has it, does he realize what an important ingredient for his palate and digestion is plain, ordinary salt, necessary alike for man and beast. He then recalls that the salt licks and salt springs have from the earliest times been centers of interest and development.

The quest of the writer for data on this fascinating subject has taken her far afield, not only into the archives of most of the states of the old confederacy, but also into departments of knowledge apparently remote from her own, so that she has had to delve into geological libraries, consult colleagues in the fields of chemistry and physiology, and do more sums in arithmetic than have fallen to her lot since she was in the primary school. But it has been a pleasant and profitable search.

The writer asks that the work be judged, not by what the individual reader may conceive should be the scope of such a study, but by the object which the author has set herself. Her aim was to make an exhaustive study of the rôle which salt played in the drama of the War Between the States so that this particular task would not need to be done again.

She wishes to express to the guardians of the archives in the many states where she has worked her warm appreciation for their generosity in placing materials at her disposal and for their constant, unwearying helpfulness. To each of the following persons she owes expression of gratitude: Mr. Robert J. Usher, Librarian of the Howard Memorial Library, New Orleans; Dr. A. R. Newsome, Secretary, The North Carolina Historical Commission;

Dr. Dunbar Rowland, Director, Mississippi State Department of Archives and History; Mrs. Thomas Owen, Director, and Mr. Peter A. Brannon, Assistant Director, Department of Archives and History of the State of Alabama; Dr. H. R. McIlwaine, Librarian, Virginia State Library; Miss Mildred Ham of the Georgia Department of Archives and History; Mr. Robert Glenk and Miss Cerf of the Louisiana State Museum for generous search for materials; and members of the staff in the office of the Adjutant General of the United States War Department. The following have been helpful by answering inquiries by letter: Mr. George C. Branner, Arkansas Geological Survey; Brigadier General Charles H. Cox, Adjutant General of the State of Georgia; Mr. G. P. Whittington of Alexandria, Louisiana; Mr. Edward Parsons, of the Louisiana Historical Society; Monsieur L. Mennig, Secretary, Société Géologique de France; and Mrs. Pearl W. Kelley, Department of Education, Division of Library and Archives, Tennessee. Expression of obligation is also made to Mr. T. L. Head of Montgomery, Alabama, and to Mr. M. M. Matthews for permission to cite their manuscript works on the salt activities of Clarke County, Alabama.

It is a pleasure to express my appreciation for the courtesy of the McKinley Publishing Company in allowing me to use one of their maps to show the locations of the salt resources of the confederacy.

Finally, my colleague, Professor Mary W. Williams, of Goucher College, has read and made suggestions for Chapter XIII, the most difficult of all the chapters of the book.

*Baltimore, 1933* E. L.

# CONTENTS

## CHAPTER I

# SALT A PRIME NECESSITY

AN EX-CONFEDERATE officer was giving a lecture in Syracuse, New York, a few years after the close of the Civil War. In the course of the day he had been driven about the city by his host and had, naturally, been given a view of the extensive salt wells and salt works for which that city was noted. In opening his lecture that evening he startled his audience with the somewhat remarkable query, "Do you know why you northerners whipped us southerners?" On the surprised ears of his audience fell the terse answer, "Because you had salt."

An inadequate amount of salt in the ration means the absence of hydrochloric acid in the gastric juice to further digestion, and a lack of sodium chloride in the blood. Its very absence as a condiment often turns what would otherwise be a delectable dish into an unwholesome one without savor.[1] Animals have been known to pine and sicken for the want of salt. A hoof and tongue disease which appeared among the cavalry horses of General Lee's army in 1862 was attributed to the lack of salt or to the use of new corn.[2] When the ration was reduced to its lowest dimensions, salt, along with coffee, sugar, and hard bread, still usually held its place as one of the indispensable articles for flavoring the meat which officers hoped to add on the way to the battle-field.[3]

According to the adjutant general of Alabama, the confederacy required 6,000,000 bushels of salt a year, or 300,000,000 pounds, computed on a reduced war basis rather than the fifty pound per capita basis which had marked the ante-bellum consumption. It was his belief that only an enormous price or a great scarcity would force the people to turn to making it,[4] for only a very small part of that amount had been produced in the southern states in the decade before the Civil War.

Until the reader pauses to ponder the various uses of salt, it is difficult to understand how such a stupendous amount could have been required for a population of only 9,000,000 people. It had its place in every ration of the soldier, whether directly by being listed in the regular items along with meal, rice, molasses, sugar, vinegar, and bacon, or indirectly when the ration was reduced to one-half pound of bacon or salt pork with one pint of corn-meal. The relation maintained between this condiment and the other parts of the ration is indicated in the following monthly allowance for the confederate soldier in 1864: 10 pounds of bacon, 26 pounds of coarse meal, 7 pounds of flour or hard biscuit, 3 pounds of rice, and 1½ pounds of salt—with vegetables in season.[5] The allowance per person for the civilian population in Virginia in 1862 was slightly less—one pound per capita per month, though the average yearly amount required for all purposes was established at thirty pounds to the head.[6]

This allowance, it should be noted, was far lower than the average yearly per capita consumption before the war. The people of the United States, according to the geologist Thomassy, enjoyed the distinction of consuming more salt than the people of any other country, as the average yearly consumption was one bushel or fifty pounds per capita. The further striking facts appear that although there was no country on the globe richer in salines than the United States of America, it imported no less than 12,000,000 bushels annually and that a striking proportion of this amount went into the southern states. The salt was imported from England and the West Indies, and a fourth of all the salt entering from England was brought to the port of New Orleans. During the three-year period, from 1857 to 1860, there entered this single port a yearly average of about 700,000 sacks or about 2,100,000 bushels of coarse salt. Allowing 30,000 tons for the other southern ports, with some deductions for variations, a fair computation shows a total of all grades of salt of 100,000 tons or 350 tons a day being discharged at the Louisiana port. England enjoyed the advantage that she could ship salt almost without expense as ballast in the vessels coming to the southern states for cotton.[7]

The French geologist, Thomassy, writing in 1860, predicted, almost with prophetic vision, the results in the matter of salt of a blockade of the south, though he was thinking of this result in the event of a war between Great Britain and the United States. "The question of salt", he wrote, "as with the majority of vital questions, has been thoroughly comprehended only under the pressure of want. When we are deprived of air, we exert all our strength to have it and fill the lungs with it * * *; in the same way, when we find ourselves deprived of salt, we make the greatest sacrifices to procure some, if it be only a few grains. Then we realize very fully the rôle of this element so necessary for health and we are convinced of its value for the well-being of nations as for the life of the individual". He pointed to the lack of salt felt in the American Revolution and in the War of 1812. He further declared that Louisiana could become one of the greatest of the states by producing salt, and that she could add a new element of wealth to rival her sugar and cotton. "Just as this element of future prosperity, this vital food, almost as necessary to their economic independence (referring to the southern states) as gunpowder has been to their national independence, is furnished them exclusively by strangers, and is found in hands which, in spite of all the dreams of perpetual peace, could easily become some day those of an enemy and be made into an instrument, if not of domination, at least of famine and internal trouble * * * so foreign salt seems to exercise the same charm for the United States as sugar for a child, particularly in the southern states, which do not yet dream of extracting it from the salt water by means of atmospheric evaporation. Thus England pours it out at her ease and abundantly, one might say expressly to dare them to the idea of utilizing their brilliant sun and the sea which kisses their shores." [8]

The allowance cited above of something over a pound a month, if generally applied throughout the states of the confederacy, would account for the consumption of only about 110,500,000 pounds.[9] One is forced to pause before he can explain the other 189,500,000 pounds of salt. There instantly flashes across the mind the thought of that great staple of the south—bacon. However, we must recall in these modern days of refrigeration and canning that smoking and salting were the only means known to preserve meat of any kind. In the packing season of 1860-61 upward of 3,000,000 head of hogs were cured at the various packing-houses of the United States in addition to those put up by farmers in local smoke-houses, an amount which cannot be calculated but which was manifestly large. It was estimated that the requirement of the confederate army was not less than 500,000 hogs annually, though it should be noted in passing that one official did not hope to secure all that amount within the limits of the confederacy.[10] To this estimate must be added the very much larger consumption of hogs by the civilian population, which numbered more than 8,500,000. Pork, beef, and mutton, including all parts of the animal, were cured or packed in brine.[11] Nearly all smoke-houses contained several large troughs, hollowed out from the stocks of trees, sometimes six to eight feet long, to hold salted pork, salt, and dried bacon packed down in leached ashes.[12] The amount of meat indicated in a letter of a North Carolina planter as so cured is not at all unusual, "I have put down in salt 4,000 lbs. of pork my own make here. It will hurry you to beat it I expect".[13] The packing season ran usually through December to February and accentuated the demand for salt during and just before that period.

There was, naturally, a well established rule among experts for the proportion of salt required to cure a given amount of meat, and the amount was large. The accepted rule was two bushels of salt to the thousand pounds of pork, and a bushel and a quarter to each beef of five hundred pounds. A lesser amount was regarded as endangering the preservation of the meat.[14]

This one, universal method of preserving meat was applied to fish of all kinds. The North Carolina fisheries alone yielded annually millions of herring and shad to be salted. On one occasion during the war when the meat supply needed to be supplemented, it was suggested that a detail of 10,000 soldiers be set to catch fish, which were to be scaled, dressed, and salted by women, white or black.[15]

In the absence of a universal system of refrigeration, it was particularly important that butter should be salted freely in order to dry it thoroughly, as absorption of moisture was recognized as the cause of quick deterioration of this delicate article. Salt was also used to pack eggs in lieu of a better agency for excluding the air and preventing them from touching each other.

The need for salt, as has been indicated, was not confined to man. Cattle in inland countries, removed from the influence of salt air, must have it added to the ration or suffer cruelly for the want of it. They will, as is well known to ranchers, expose themselves to great danger in order to secure it. Concern for this item for the animals in army use is reflected in such commands as, "Cut and cure corn for your horses, salting your animals well—Do not let your supply of salt for the horses run short—" "Supply your command with forage by cutting corn from the fields and partially drying it in the sun, by cutting it up and feeding it with salt for your stock." [16] Inferior hay was sometimes camouflaged for the poor dumb beasts by sprinkling it with brine.[17]

The hides of slaughtered animals, precious for harness as well as for shoes, had to be preserved for leather by means of salt. The modern reader, thinking of war in terms of automobiles, aeroplanes, and tanks, has to be reminded of the enormous numbers of horses used for the cavalry and artillery and of the thousands of mules for the supply wagons. All of the planters were reduced to the necessity of tanning leather for their own use and also for supplementing the supply available for soldiers' shoes. The hides of horses, mules, hogs, dogs—all were used, and the only agency known to keep the hides from spoiling until time was afforded for tanning them was salt sprinkled heavily upon and between them. Small quantities of salt were also called for when the plantation mistress resorted to dyeing her homespuns and wished to set the color.[18] To a limited degree the chemical properties of salt as a fertilizing agency were believed by some planters and scientists to improve the soil.[19] It was held useful, for instance, in improving sandy pine lands and in killing the roots of some weeds. Indeed, it was said to be the only remedy to extirpate the weed known as *compositae*. The natural manuring of the cotton plant with salt from the ocean was held to explain the superiority of the sea-island staple.

The constant use of salt meat, however, was not an unmitigated good for the army. A diet too heavy in salt, unless relieved by fresh green vegetables and fruits, produced scurvy, while the lemon and potato, main dependencies of the physicians as anti-scorbutics, were sadly lacking in the confederacy.[20]

## CHAPTER II

# SALT RESOURCES WITHIN THE CONFEDERACY

AT THE outbreak of the Civil War there were three methods of producing salt known and practised: extracting the salt from saline artesian wells, boiling down sea water or water from inland salt lakes, and mining deposits of rock salt, formed by gradual evaporation of bodies of sea water which had by geological process been cut off from the ocean.[1]

In early ante-bellum days when salt was needed on a plantation in the Black Belt of Georgia, Alabama, and Florida, the overseer with pots, kettles, and slaves would go to the nearest salt wells in the neighborhood or to some suitable location on the extensive seaboard and camp out for several weeks in order to make the year's supply of salt. When the blockade shut off the usual sources of supply early in the Civil War, the planters reverted to pioneer methods and again visited salt licks, salt wells, and the seacoast.

While it is possible to locate salt springs or salt lakes in nearly every state of the confederacy, certain salines were of such outstanding value as to be of importance for the entire south. Significant examples were the following: those on the Great Kanawha River, three miles above Charleston, now in West Virginia; the Goose Creek Salt Works, five in number, all privately owned or rented, which were situated either on the main fork or on the branches of that stream within a radius of five miles from Manchester, Kentucky;[2] the wells on the state reservation of Alabama in Clarke County, which, together with the privately owned wells in Washington and Mobile Counties, supplied the interior of Alabama, Mississippi, and Georgia from 1862 through the rest of the war; the salines of north Louisiana; and above all the great wells in southwestern Virginia at Saltville in Smyth and Washington Counties, which lacked only labor to supply the whole confederacy. The Goose Creek Works were lost by the confederacy so early in the war, as will be related later, that space will not be devoted to a detailed description of these wells.

The Kanawha Licks had been located and extensively worked from the first settlement of the valley, and from time immemorial had been frequented by Indians and swarms of buffalo, elk, deer, and other wild animals. The tradition of how the treasure was discovered is too fascinating a tale to be omitted. Two boys, on one occasion when the water was low, went fishing. On the return journey, they observed a spring of pure, sparkling water, bubbling up close to the edge of the current; to slake their thirst, they filled an empty milk bottle from the spring, only to find it to their utter astonishment salt water. News of the strange incident spread; experiments by boring soon revealed an abundance of brine. The first furnace was erected in 1797. As no means were used to purify the brine of its carbonate of iron, the salt when crystallized had a reddish tint, to which was ignorantly attributed the strength for which Kanawha salt soon became famous. From far and near came orders for that "strong red salt from the Kanawha Licks." After 1808 salt manufacture became one of the leading industries of the region and it expanded as new improvements were introduced: coal for fuel, a steam engine to pump up the brine, and purification of the iron from the water. Just prior to the outbreak of the war its production was 2,500,000 bushels a year.[3]

The salt springs in Alabama had been discovered through some Indian tradition by a Scotchman named McFarland, who about 1809 opened what were always locally known as the Lower Works. By congressional act in 1819 the public lands on which the salt springs were situated were donated to the state with stipulations as to the conditions on which the springs might be leased or worked. The state legislature thereupon promptly authorized the governor to lease the springs on terms to insure their being worked most extensively and "most advantageously to the state." All private makers were required, accordingly, to give a certain amount of the salt made to the state. Three localities in Clarke County, in the southwestern part of the state, became differentiated as the best-known locales for salt in the state: the Lower Works, just north of Oven Bluff, the Central Salt Works, near Salt Mountain, and the Upper Salt Works, range one of Township Seven.[4] The state reservations were at the Upper and Lower Works. There were also many smaller works scattered around in Washington and Mobile Counties, as well as in Clarke County, where salt was made for domestic use, for neighboring plantations, and for sale.

At this type of works wells were usually bored to depths ranging from sixty to a hundred feet—a few even to six hundred feet—while in bottom or swamp lands, such as characterized the salt lands in Alabama, brine might be encountered at a depth of only eight feet, though the average well was much deeper. When the wells were first sunk here, the water rose to the surface and overflowed at the rate of thirty gallons a minute, but, as the number of wells multiplied, the brine ceased to flow as an artesian well so that it had to be pumped about sixteen feet by steam or horse power. It was then boiled in large iron kettles much like those used for sugar, and salt secured in much the proportion of cane sugar—one kettle of salt to seven or eight of brine. The salt-making season was limited to the months of April to December, for, when the Tombigbee River was full and overflowing its banks, all of the salt lands were submerged, and any work whatsoever thus prevented. Even after the water drained off, the seepage of fresh water into the wells was too great to permit of salt-making for several weeks.[5]

The salt licks in northwestern Louisiana, which had in the early years of settlement been the scene of annual gatherings of planters to secure the needed store of salt, were destined to resume during the war their lost importance. Neighborhood delegations, commercial companies, and families were to flock to them during the years 1861-65 as they had in the decade of 1830-40, so that they became the scenes of great activity despite the weakness of the brine procurable at most of these wells.[6]

Most of the salines were areas where brine issued in crude fashion and evaporated, indicating bodies of rock salt below the surface. These licks, not infrequently supplemented by salt springs, occur sporadically scattered north of Red River, chiefly on Saline and Dugdemona Bayous [7] and on Lake Bisteneau in Bossier, Bienville, and Winn Parishes. They presented a remarkable appearance. The face of the country is moderately hilly, and moderately heavily wooded with a growth of thorn and long-leaf pine. The licks usually lie in a broad, flat-bottomed valley, rarely more than 160 feet at the very most above sea level, resembling the bottom of a dried-up pond; they are, indeed, located in the basins of ancient lakes. The bottoms are usually low and swampy, heavily wooded with swamp palmetto, gum, oak, and an occasional cypress, except where the brine approaches the surface and produces barren lick spots. There the ground is covered with a saline efflorescence as white as snow.

The best known and largest of the works were King's Salt Works; Drake's on Saline Bayou, second in size, with very pure brine almost free from gypsum; Rayburn's in the southeastern part of Bienville Parish; Price's, which never produced a great deal of salt; Bayou Castor Saline and the various licks near Rochelle; Cedar Lick; and finally those on Lake Bisteneau, the largest by far in the section.

At some of the licks broken fragments of pottery and

Indian arrowheads seem to indicate temporary campsites of the aborigines, established obviously for the purpose of obtaining salt. Especially do the Indians seem to have frequented Little Lick near Saline Bayou, judging from the large accumulation of pot fragments and from the fact that the pottery appears to have been made on the spot. A reference to the procuring of salt at this place may be found in historical records as early as March 22, 1700, in the *Journal* of Bienville, who had noted Indians resorting there for salt.[8] Evidence of Indian occupation may also be found in the Lake Bisteneau region to the south and east of Potter's Pond, which owes its name to the accumulation of pot fragments near it.[9]

Accounts of the making of salt at the Louisiana salines by white men at an early period are not lacking. One of the first occurs in a letter by a certain John Sibley to General Dearborn in 1805, when the former noted two old crippled men making salt for the whole district at Little Lick on Saline Bayou, a location which became known later as Drake's Salt Works. Twelve saline springs were open there at that time. By the year 1812, only seven years later, three wells had been sunk near Natchitoches.[10]

The local demand for salt became so great that Reuben Drake in the early 40's tried to secure a stronger brine by deep-boring. Eight wells were bored, in each of which artesian brine was found. The pressure in one well having a depth of 1,011 feet was found sufficient to raise the water into a tank thirty-five feet above the opening of the pipe, while it furnished a flow of eighteen to twenty gallons per minute. This well was not, however, greatly used, as the brine was much weaker than in the shallow wells and contained about two per cent of solid matter.[11]

The main lick at the group of wells known as Rayburn's, located a mile beyond the Saline River, was one of the most extensive. It lay in a flat, circular, slightly swampy area of forty to fifty acres, fringed by a few stunted hawthorn trees which merged into the surrounding forest. The southern end was so swampy that in heavy rains it was flooded to a depth of two or three feet, a condition necessitating the protection of the wells by low levees. Not till the early 40's, when the owner of the land began operations on a modest scale, was salt made here regularly. The water was rather stronger than at some of the licks, yielding salt in the proportion of one part of salt to six of water.[12]

At Price's Lick, which lay southeast of Rayburn's, the licks and salt works extended in a circle around Lick Hill, and acquired their reputation only in the Civil War. The location, however, was noted on Tanner's map of Louisiana as early as 1820, marked *Salt Lick*. The exact date when the first effort at salt-making was made at that place is not known, but John Walker, who moved there in 1859, found old wells on the lick, thus proving that it was known before the outbreak of hostilities in 1861 focused attention on all sources of salt supply. The water, when tested, showed about one bushel of salt to eight of brine and would hardly float an egg.[13]

King's Salt Works were erected fourteen miles west of Rayburn's, close to a rather extensive outcropping in a level hummock bordering Bayou Castor, a tributary of the Black Lake Bayou. It ranks fifth in size among the north Louisiana salines. Although King owned only a little well beyond the main lick, his name continued to designate the entire group which grew up near his works during the war. Salt operations were begun by King about the time of those at Rayburn's in the 40's. After digging a shallow well, he drilled into the solid rock to a depth of 136 feet; the water then rose to within two feet of the surface so that it could be reached by a crude pump. He constructed a rude salt-house, and every year after the crops had been garnered, surrounding planters repaired with their negroes to the well to produce the requisite supply of salt for the ensuing winter, even though the brine was not above the strength of sea water.[14]

Far and away the most important of the salt works in this region, both in point of size and of strength, were those located on the eastern shore and at the head of Lake Bisteneau, which is one of a series of lakes formed along the Red River by deposits of silt and the consequent choking of the outlet of tributary streams. In the old bottom of the lake, which was bare of trees at the time of the war, are a number of elevated areas, resembling islands and so-named, which rise ten to fifteen feet above the surrounding bottom. Exploitation of the natural resource dates from 1846, when B. M. Thompson and W. S. Howard carried samples of brine from Potter's Pond in this area to their homes in order to test its strength by boiling. The quality indicated evidently proved disappointing, for it was not until some years had elapsed that salt was manufactured here in any quantity. After 1850 two brothers by the name of Hodges made a little salt. The year 1855 was memorable for a drought so severe that boats could not come up the lake from the settlements on the Red River with the usual supplies. The salt reserves without importation ran so short that the entire neighborhood betook themselves to this well-known lick to make their salt. When rains opened up the navigation, salt-making at Bisteneau again almost ceased.[15] There were several other salines in the general region such as Castor Salt Springs, which had been developed before 1860 [16] or were promptly exploited when the war forced the southerners to utilize their every resource.[17]

But the salt resources of Alabama and of north Louisiana fade into insignificance beside those of Virginia, on which in large measure the entire confederacy east of the Mississippi depended after the loss of the Kanawha wells. Indeed, for more than one hundred years they had been the source of supply for southwestern Virginia, eastern Tennessee, eastern North Carolina, and for parts of Georgia and Alabama, until they could no longer compete with the Kanawha product.

The wells were concentrated in Washington and Smyth Counties [18] in the extreme southwestern corner of the Old

Dominion. Saltville, the point where most of the manufacture of salt was located, marked the southern limit of the extraordinary deposits of salt which trace for seventeen miles the line of a great fissure in the crust of the earth along the North Fork of the Holston River. Only that singularly beautiful limestone basin encircling Saltville yielded much salt. It is now established that a deposit of solid rock salt more than 175 feet thick exists below the surface of the earth at a depth varying from 200 feet. When it is realized that there are probably 500 acres of land underlaid with rock salt, it can be readily understood why these deposits could be drawn upon so long and so heavily without losing any of their original strength and why it was currently stated during the Civil War that Saltville lacked only labor adequately to supply the entire confederacy with salt. Wells were bored at intervals from 1773 to 1854, most of them about 1847,[19] the most important of which were known as King's and Preston's Wells.

Three hundred and thirty acres, embracing the area to be made famous by the salt lick in Smyth County, was patented to Charles Campbell in October, 1753.[20] At that date one-third of the area was covered by water, and naturally named Salt Lake, while another third was a morass of flat marsh land overlying a brittle limestone, which was drained later by a channel to the Holston River. It was a favorite resort of buffalo, elk, and deer; fragments of shells and pottery bear witness to the fact that the Indians erected wigwams near the mysterious lake and presumably made salt from the springs which rose to the surface at the eastern margin of the valley and at the edge of the creek. The clear, bright crystals, forming during the warm summer days on the limestone rocks over which some of the springs flowed, pointed the way to the native in search of salt.

The property remained undeveloped until about 1782,[21] perhaps until even a few years later, when Colonel Arthur Campbell, executor of General William Campbell's estate,[22] finding that some persons had made salt at the lick, decided to develop it. Foreseeing the need of a large supply of fuel, he entered several tracts of land adjoining and near the salt lick tract in the name of the infant son of General William Campbell. The boy's early death left the property to his sister, Sarah Buchanan Campbell.

A family dispute resulted in a new guardian, who came to the lick about 1790, dug a well on the edge of the flat, built a log salt-house, and began the manufacture of salt after the primitive methods of the time. The salt was lifted from the kettles by long-handled dippers, put into baskets of splints over the kettles to drain, and then emptied into the salt-house, which was located at the end of a platform some thirty feet from the furnace.[23]

General Francis Preston, who had married Sarah Buchanan, now enters the story by locating his house on the grounds and embarking on the manufacture of salt, which was developing into an important industry. The brine rose in the wells to within twenty feet of the surface, from which depth it was drawn up in large buckets by a windlass and emptied into an open trough which conveyed the water to cisterns near the furnaces. Since the surface and seepage water drained into these open shafts, the strength of the brine rapidly diminished, a fact which necessitated the digging of other wells.[24] These salines, Preston's, were on the north fork of the Holston River, about seventeen miles east of Abingdon and half a mile south of the river.

In 1795 William King stepped upon the stage by purchasing a tract of 150 acres adjoining the lick on the west. He offered to sell his land to General Preston, but the latter felt that there was salt enough for both of them and preferred King to any other rival. It is not known exactly when the latter began his well, but in 1800 he had struck a saturation of salt water at 200 feet, had constructed furnaces, and was boiling brine heavy enough to yield a bushel of salt from 32 gallons of water.[25] By 1805 as many as 200 bushels of salt, declared to be the equal in quality of Liverpool salt, were being made in a single day.[26]

In 1801 began the series of leases into which it is not necessary to enter, when King rented the Preston salines, together with many privileges, for a period of ten years for the goodly sum of $12,000 a year, payable in cash, salt, and merchandise.[27] It may suffice here to say that from this time the policy of the lessees was to prevent competition, so that when William C. Preston bought up the neglected furnaces on that estate until there was formidable competition, the lessee of the King wells felt that a lease of the Preston estate was obligatory. Efforts to unite the two estates under a common management in 1845 failed. With much litigation and many leases of the Preston wells to King and of the King estate to a member of the Preston family and of both estates to an outsider, they jogged along until at the outbreak of the Civil War a firm by the name of Stuart, Buchanan and Company held control of the Preston property and of three of the King wells by a lease. In early June, 1862, this firm became owners of the Preston estate. This exact status of part-owner, part-lessee, must be kept in mind in order to understand the complications which ensued in connection with the preemption of the wells by Virginia in 1864.[28] Already in 1860 a large number of wells had been sunk in the valley area, five of them put in active operation, representing a value in investment of capital, annual cost of labor, and raw materials of $83,000 with an annual output worth $72,000.[29] The brine was strong, so strong indeed that the water from some of the wells would not dissolve a particle of salt. In chemical analysis it proved very pure, showing from 98.39% to 99% of sodium chloride with less than one-half per cent each of sulphate and iron.[30] It therefore ranks higher in purity than the Kanawha salt, which is only 79.45% sodium chloride, and the saturation is heavy, as eighteen gallons of brine yield one bushel of salt.

The arid alkali section of western Arkansas, Texas, and

what was then known as Indian Territory, afforded some opportunity for supplying, locally at least, the need of salt from salt springs. References during the war prove that such sources were found and utilized in Arkansas: one spring not far from Arkadelphia; some on the White River; some near the Louisiana state line; and some on the Ouachita River, where a generous supply of brine brought vigorous operations in 1862. The report of the destruction in August, 1864, of small works near Little Rock, though these produced only two bushels a day, indicated a sorely needed source for local needs, since they were alluded to in a report as "the only works about here." [31] A geological survey made in 1859-60 shows two locations in Sevier County near the boundary line between Arkansas and Indian Territory, at one of which twenty bushels of dry, white salt were at one time made daily. Under war exigencies they would naturally be made again to produce to their full capacity.[32] A map of Arkansas, compiled in 1863 from the United States survey and from a letter written in 1845, in which the location of the salt springs granted by congress to the state is specified with great precision, shows saline springs at eleven different localities.[33] A report, written early in 1861 to Governor Clark of Texas concerning the lands of the Indian tribes, whom the confederates were industriously courting at that time, reveals the presence of "fine salt springs and wells" in that territory. A military officer held in the fall of 1862 that he could manufacture all the salt he needed within the limits of the Cherokee nation.[34]

Confederate authorities looked to Texas for a partial supply for the Trans-Mississippi Department. Steen and Brooks salines, in the northern and southern parts of Smith County respectively, despite the weakness of their brines, were available for Civil War exploitation. Salt was also made from brine at Grand Saline in Van Zandt County in 1860 in utter ignorance that there lay under ground a huge deposit of solid rock salt. A geological survey of this state, published in 1866, must obviously be describing salt works which were in operation during the war. It locates works in two counties in central Texas, Llano and Lampasas, the latter called in the records of the time, Lamposso. Below the mouth of Falls Creek in Llano County at the base of a ridge salt works produced some twenty to thirty bushels a day, while about twelve miles below, lying in a picturesque gorge in the neighboring county not far from the Colorado River, were located Swenson's Salines, which yielded about a bushel of salt water a minute. To concentrate the salt from a weak brine a process of reduction was employed here which the writer has not encountered elsewhere in this study, although it was in use in Germany and France with weak brines. The water, after being pumped into a trough erected on a scaffold forty feet high, was scattered over cedar boughs spread over the scaffold, in order to increase the evaporation so that the water might fall into the vats below in a concentrated brine strong enough to boil.[35]

The second process, that of boiling down salt water from the ocean or from salt lakes, is distinguished from the first merely by the fact that the water is available without sinking wells. While salt-making went on along the entire Atlantic and Gulf coast,[36] no state played so large a share in this enterprise as Florida. Her immense seacoast of about a thousand miles, her many secluded bays and inlets, her comparative isolation, her poverty, her cheap fuel in the form of uncleared forests and brush, and, last but far from least, brine available without the labor and expense of pumps or wells, were assets of great value. They made her one of the most important states to the confederacy for the manufacture of salt.[37]

The hard-pressed confederacy could not overlook the dead lakes in western and southern Texas as possible sources of supply. These logically classify with the Atlantic and Gulf coasts, since the salty deposit was merely scraped from the surface in crystallized form instead of being dipped from the sea to be boiled into crystals.[38]

Many of the salines of this description were low, flat, marshy areas, holding water during the winter, but acquiring an incrustation of salt when the summer sun evaporated the water. In several of these so-called lakes in the arid El Paso County the valuable salt deposit on the surface occasionally attained a depth of an inch or more. As the ground-water level is very near the surface there, when one layer of salt was removed, it was quickly replaced by brine which formed a fresh deposit in a few weeks. Mexicans had long been in the habit of coming across the Rio Grande to this salt basin from Chihuahua. The first wagon road to the deposit was not built until 1863, in the midst of the war, whereupon the area was claimed and an effort made to institute charges for a privilege which had hitherto been free. From a distance these so-called lakes look like ponds covered with snow, though during the dry season the surface becomes coated with impurities. Close at hand the salt is grayish white. Such incrustations of saline matter also formed for miles along the forks of the Brazos River and along the Salt Fork of the Red River in northwestern Texas, when the water was at low stage so that its banks appeared as white as snow. The least evaporation of the water along the margin of the river resulted in the precipitation of salt.[39]

Certainly the most interesting lake of this type was *El Sol del Rey* in Hidalgo County, close to the Mexican border, which, reserved originally as the property of the crown of Spain, had passed to Texas as an unpatentable possession, and so became at a critical juncture a source of revenue and supply to the state and people of Texas.[40] Such natural stores would not wait upon the weather, for heavy rains dissolved millions of bushels of salt formed in these "dry" lakes.

Fortunately, statistics afford us a very fair knowledge of the amount of salt derived from the various salines of the southern states. In 1858 Virginia, Kentucky, Florida, and Texas yielded 2,365,000 bushels as compared with 12,000,000 bushels derived from New York, Ohio, and Pennsylvania.[41]

What must have been regarded at the time of its discovery as a pure gift from Heaven was a salt-mine on Avery's Island, which is located ten miles southwest of New Iberia in southern Louisiana, designated an island by virtue of the surrounding bayous and marshes. The existence of salt wells on Petite Anse was widely known. Salt had first been discovered as early as 1791 by a John Hayes, who one year after settling there had, while hunting, discovered a brine spring in a ravine near the center of the island. Not long afterwards Jesse McCaul bought nineteen acres of land, including the salt spring, and began to make salt, but the supply was too limited to pay the expenses of operation. Therefore, after digging several wells, he abandoned the enterprise. The springs were then neglected until during the War of 1812 when the price of salt rose to such a height that John Marsh renewed the production for the brief period of that war and supplied to a large extent the demands of Attakapas and of Opelousas Parishes.

Judge D. D. Avery, owner of the island at the outbreak of the Civil War, when the blockade made the product scarce, began to work the spring systematically. His son, John Avery, to increase the supply of brine, undertook to clean and deepen the well. On the fourth of May, 1862, the negro at the bottom of the hole, then sixteen feet deep, found, as he thought, a log which he could not remove. It proved to be a bed of rock salt, clear as crystal. The surface of the rock mass is irregular so that its distance from the surface of the earth varies, but it rises to a few feet above tide level, a fact which explains the presence of the brine springs known on the island prior to 1862.[42] The mass, thought to be at least forty feet thick, was estimated at 7,000,000 tons of workable material, an estimate found later to have been altogether too modest,[43] since the mass was soon known to cover 144 acres. A great deposit of fragments of clay pots and ashes in places three to six inches thick and extending over an area of possibly five acres testifies to the extent of salt operations in prehistoric times. Human handiwork has been found so close to the surface of the salt bed as to indicate that the mass of rock salt was once known, though the boiling process alone has been resorted to in traditional times. The discovery of rock salt at that point was a great surprise to geologists, with the exception of Thomassy. It was found on one of a chain of five islands rising partly from the sea, partly from the coast marshes between the mouths of the Atchafalaya and Vermilion Rivers.[44] Its value was greatly enhanced by the fact that it proved to be perfectly dry, homogeneous, and remarkably pure, though the reverse condition is usually true of rock salt. When quarried, as it was during the Civil War from large open pits, it comes out in huge blocks of ice-clear transparency. It is so hard that it has to be blasted with dynamite.[45] This important discovery led some wag to say that on this plantation sugar could be grown above and salt below the ground.[46]

The approach to the mine was over a causeway two miles in length, built during the war over the bayou and marshes from New Iberia. By September, 1862, only one shaft had been sunk to a depth of about twenty feet from the surface, and the rock salt blasted to a depth of fourteen feet, though a second shaft was being sunk at that time. Exploratory tunnels in two directions from the shaft revealed the continuance of the salt-bed at no point more than thirty feet below the surface of the earth.[47]

The desire for such a wind-fall in other localities gave rise to newspaper rumors which duplicated the mine at several other points.[48] As soon as the reports concerning the New Iberia mine were authenticated, citizens and governors of neighboring states were besieging Judge Avery for contracts, as will be detailed later in this story.[49] There may have been, as Governor Pettus wrote President Davis, "salt for all the Confederacy" in there,[50] but the prospects for getting it out, except by government seizure of the works, seemed slight. Transportation to the Cis-Mississippi region,[51] through the Teche to the Atchafalaya, thence by portage to Alexandria and then down the Red River to the Mississippi, circuitous at best, was cut off by the fall of Vicksburg in July, 1863.[52]

## CHAPTER III

# PRESSING SCARCITY OF SALT

THE PRIVATIONS endured during the life of the confederacy are a matter of tradition, but can perhaps be best appreciated when the lack of such a simple commodity as salt is closely studied. The fundamental causes, as in the case of all other articles of which there was a scarcity, were threefold: heartless extortion, inadequate transportation, and the blockade—with emphasis on the last factor, since domestic sources of supply had to be found for thousands of pounds which had formerly come from the north or from Europe.

As early as the fall of 1861, when the half-million sacks which had entered New Orleans in 1860-61 had shrunk to almost nothing,[1] stringency in the supply of salt was beginning to be felt and anxiety for the future supply manifested. A complaint to Governor Pettus is typical, not only for Mississippi, but for all the interior and lower states. "With a great many now, the deepest anxiety prevails to keep our families from suffering for want of salted provisions. Meat is now ready to be slaughtered." [2]

Already the twin evils of speculation and extortion had raised their heads, and continued to run riot through most of the period of the war. Men resorted to every known trick and fraud to control the supply of the commodity in defiance of the thunders of pulpit and press and despite executive interference. "The demon of speculation and extortion," in the vehement language of one irate governor, seemed to seize upon nearly all sorts and conditions of men, until all necessities were beyond the reach of the poor. Speculators swarmed everywhere in the land, offering fabulous prices for everything and representing themselves under forged orders as agents of the government. One indignant Mississippi planter sought to spur the governor to action by writing, "If the strong arm of the law be laid upon one class of harpies, why not exterminate every class of them? I can see no difference. A law should be passed in every state prohibiting the planter from selling to such men." [3] An editor wrote less belligerently, "It is currently reported that all the salt in New Orleans and elsewhere is now in the hands of speculators * * * The planters are ready to put up four times the amount of meat this winter that has been put up by them heretofore, but they cannot do it without salt, and unless they can get it soon, they must suffer in consequence * * * Something must be done in the matter, and be done quickly. We are willing that speculators should reap a rich profit, but we are not willing for them to suck the very life blood out of the people, if we can avoid it." [4] One planter wrote semi-jocosely to Governor Vance, after recording the price of salt at eight dollars a bushel, "Blessed are they that have no hogs. these extortioners are worse than the Hessians & doing our cause more damage." Six months later he reported, "My dependence is upon blackberries." [5]

By October Governor Moore of Alabama deemed it his duty to protest in a public proclamation "against such conduct, and pronounce it unpatriotic and wicked; and I hereby notify all persons authorized to make purchases for the state of Alabama, not, under any circumstances, to buy at the unreasonable prices which may be exacted by such persons." [6]

One man succeeded in getting a "corner" on all available salt in West Florida in 1861 and in holding it for high prices. Refusal to sell at a reasonable price produced excitement which threatened to end in a riot with seizure of the salt by a mob. As Apalachicola was threatened by the enemy and the salt liable to fall into hostile hands at a time when that commodity was greatly needed by the army, it was removed ninety miles north for storage and distribution by sale.[7] Governor Milton of Florida accordingly complained of the vile spirit of speculation and extortion.[8]

Governor Brown stigmatized the price in Georgia as "enormous and exorbitant", owing to speculation and extreme scarcity.[9] He fought valiantly against monopoly of the coast lands which bordered on valuable and accessible localities for good salt water, securing from the legislature an act prohibiting the rental or leasing of more coast than could be used for salt manufacturing, and holding renters to actual production within six months from the date of such leasing or renting.[10] The general assembly of that state passed an act in 1861 penalizing heavily various acts of extortion, engrossing, and speculation. Governor Brown, however, wrote to the governor of Virginia in November, 1861, "While all of those unpatriotic, not to say unchristian, practises have been rife in the State, ever since the passage of the Act, it has proved almost a dead letter upon the statute-book. The Judges find it difficult to have bills of indictment found under the Act, and if a single conviction has yet been had, I have not been informed of it." [11]

Devious methods were resorted to by the extortioners. Men were glad to plead the law against exportation of salt from the state to escape from their contracts, even though they had been made and payments accepted before the passage of the act.[12] Speculators sought to fill up cars set apart for exclusive transportation of salt.[13] It would appear that salt claims, owned by citizens who had fallen within the enemy's lines, were bought up with a view to speculation.[14] One case of greed is so striking that it is worth citing. A citizen, with an eye to private gain rather than to salt supply, insisted on receiving his pay for twenty-five slaves to be hired to the superintendent of the Virginia salt works at Saltville, in good salt at two dollars a bushel instead of in money. It was obvious that he meant to sell his booty at the highest price it would command, for he was deaf to the pleas that such a quantity of salt should not be taken from the people to fall into the hands of an individual.[15]

It is difficult to convey adequately the depth of feeling which was aroused on this subject of speculation. The governors of North Carolina, Virginia, Louisiana, and Tennessee, as well as those of Georgia, Florida, and Alabama, condemned the practice in ringing tones. The *Daily Richmond Examiner* thought that those guilty of the practice could be fined and imprisoned by common law for "regrating and engrossing", and if they could be reached by neither common nor statute law, it expressed the pious hope that they could be handled by lynch law."[16] Naturally, wild rumors of profit were rife: one person was declared to be asking 2,000% profit on 10,000 sacks of salt; another in St. Louis was accused of smuggling $4,000 worth of salt into Arkansas where he realized $65,000. One woman assured Governor Vance that in the little town of Concord, North Carolina, the extortion spirit had filled large warehouses, "until there is danger of their falling from the weight of their contents." [17]

Though salt was scarce throughout the war, except possibly in Virginia in 1864, the clamor for it rose in a mounting crescendo during the first two years, probably striking the most strident note during the packing season of 1862-63. Its price, its scarcity, and the devices to reduce the amount requisite for daily use replaced the weather as topics of conversation. The offer of salt for sale could lure forth the hoarded silver as the lesser commodities, such as shoes and clothing, usually thought indispensable, could not. The mere entry of a lot of salt into a port rose to the dignity of news, while advertisements of a shipment of salt for sale were heralded with exclamation marks.[18] Planters in North Carolina complained that they had not a pound of salt; the question of the necessity of a special session of the legislature to deal with the question was raised in several states; letters to the governors from gentlemen planters down to the humblest citizens were incessant, especially in behalf of the soldiers' families; hints of the danger of mutiny were thrown out. The supply for an entire state was not sufficient to meet the demands of one county. Concerted action was taken whereby one agent was delegated by his neighbors to present for a district or county their united appeal to the governor.[19] The following appeal may perhaps be regarded as typical of those which came from the humbler element of the population: "I have tried most every Works but can't get it, and I do not believe there is a County in Mississ. suffering as mutch as Carroll. There are those too poor to sen their Waigons to the Salt Works to La., I know Soldiers Families, they have not one pound of salt in the House."[20] One member of the firm of a packing-house just west of Saltville complained to the commissary-general in December, 1861, that he did not have one day's supply of salt on hand; that he could slaughter 8,000 to 9,000 hogs a day if he were kept supplied with salt.[21] One letter found in the Vance Executive Papers suggests clearly that the limit of endurance could be reached: I. A. Reves wrote to Governor Vance on December 15, 1862, "We have a large supply of hogs in Ashe (County). If we could get salt to save them and there is thousands of bushels at Saltville, Va., and we can't get it and we the citizens of Ashe has come to the conclusion to go and take it by force if the oners of it won't let us have it for a fare price & the people of this county has Requested me to write you on the subject & we wish your advice & whether we can be seriously pusnished or not we are willing to pay a fare price for the salt, we are compeled to have salt while it can be had or we will fite for it."[22]

If the reader were to trust the statement of the Joint Committee on Salt of Virginia in regard to the situation in 1864, the period of scarcity of salt would seem to have been passed. People and the markets were "so fully supplied with a sound, merchantable article, that despite a depreciated currency and fabulous prices for all other articles of consumption, salt alone is quoted in the price lists as 'dull and flat' and 'cheap'." [23] But this evidence is not borne out elsewhere. Daniel Worth, the Commissioner of North Carolina, reported on May 6, 1864, that at least three-fourths of the private works on the coast had long since suspended on account of scarcity of supplies and labor, though the market price of salt stood at twenty-five dollars a bushel.[24] Even in 1865 prices indicate extreme scarcity, while local agents in North Carolina complained to the governor of great dearth.

Army officers everywhere during the early period were reporting their troops without salt and that article "not to be had", earnestly begging that a little be sent. The importance of securing a supply of salt to put up beef and pork was constantly before them.[25] General Pemberton, eager to cure bacon, prohibited the exportation of hogs from his department in the winter of 1862-63, but the difficulty of obtaining a sufficient quantity of salt at the proper season prevented the success of the plan.[26] Governor Pettus declared that a large amount of pork from north Mississippi had been lost to the army for the lack of that article.[27] From the Shenandoah Valley came the complaint of a deplorable condition in a cavalry brigade because the horses had to subsist on young grass without salt.[28] Naturally, the union armies felt the lack of the commodity, as they pressed into the seceded territory, so that salt was referred to by federals at times as a luxury.[29] The complaints during 1864 from military men seem to have come largely from the outlying sections, such as Indian Territory and Arkansas, but it is hardly safe to draw the conclusion from such negative evidence that the armies in the East were supplied with the commodity. It may merely indicate acceptance of scarcity of a given article as a foregone conclusion.

The dearth of the condiment among civilians was, with exceptions here and there, severe, widespread, and continuous. The state governors were genuinely troubled and anxious concerning a supply for their citizens and for the soldiers of the state line. In July, 1862, Governor Brown of Georgia wrote, "I shall do what I possibly can to supply all during the summer and fall, and I trust with what may be made by the Georgia Salt Manufacturing Company, whose office is located at Augusta, and with what our people will make upon the coast, (it is expected that all who live near the coast will at least make their own supply), that all who practise strict economy may have enough. I respectfully suggest to our people to so divide what they may receive among their neighbors, till more can be had, that none may suffer." [30]

Governor Shorter asserted of Alabama that "The danger of a salt famine is now almost certain, and there is scarcely any misfortune which can befall us which will produce such wide-spread complaint and dissatisfaction. If the destitution could be limited to people at home, who can shift for themselves, it would not be so bad, but the families of our soldiers, far away, many of them, helpless and poor, appeal to us in language which cannot fail to excite our profoundest sympathies." [31]

Governor Pettus in a letter to President Davis in October, 1862, declared that the destitution of salt was alarming, that many of the people of the state had had no meat for many weeks because they had no salt with which to season fresh meat.[32] In Virginia itself, the greatest source of salt in the confederacy, it had become impossible to secure it except at fabulous prices, and even then not in sufficient quantities to supply the needs of the people of that state.[33] There was destitution of salt as well as of clothing and corn in the interior of North Carolina,[34] while records of suffering for salt in Tennessee are to be found in union documents.[35] The presence of important natural deposits in Louisiana did not prevent the people of the parishes of that state east of the Mississippi from starving for want of salt and salt meat.[36]

Even more illuminating as to the scarcity are the comments which have come down to us in diaries of civilians.

Everywhere the salt was dealt out in the smallest quantities, rarely more than a sack to a planter, a few pounds to a city resident, and then only when it was to be used for actual home consumption. "Sometimes," writes Mrs. Avary, "we got hold of a roast, or we would buy two quarts of flour, a little dab of lard, and a few pinches of salt and treat ourselves to a loaf of bread, which the old negress cooked for us, charging ten dollars for the baking." [37] Another woman in South Carolina relates, "It happened that my host at Radcliffe, just previous to the breaking out of hostilities, had ordered a boat-load of salt, to use upon certain unsatisfactory lands, and realizing that a blockaded coast would result in a salt famine, he hoarded his supply until the time of need should come. When it became known that Senator Hammond's salt supply was available every one from far and near came asking for it. It was like going down into Egypt for corn, and the precious crystals were distributed to all who came, according to the number in each family." [38] Dining on a few potatoes minus salt was a common event. Salt rose to the dignity of being sent in small packages as highly valued gifts. It even figured as a wedding gift at General Pickett's marriage. Moreover, as the northern army took possession of the confederacy, officers took pity upon households where neither sugar, tea, nor salt was to be found and sent some from the commissary stores of the federal army.[39]

There were, of course, many instances where salt was given generously to the poor and suffering by those better situated. A case of peculiar interest was that of an officer from Yadkin County, North Carolina, who gave salt worth $3,000 to the wives of the volunteers in his company. In order to obviate the cost of hauling the commodity, he requested Governor Vance of his own state for a letter to Governor Letcher of Virginia and to the superintendent of the railroad, requesting the use of two cars from Saltville to Max Meadows to save a haul of sixty miles. He added, "It will afford me infinitely more pleasure to give it to them than to sell it to others for large prices."[40]

Under such circumstances it could hardly excite surprise that thefts of salt are recorded. A North Carolina planter lost four barrels of salt from a wagon; an Alabama salt agent lost five sacks from a train in transit, the loss of which, involving ten dollars, he tenaciously insisted should be borne by all the citizens of his county.[41] More serious was the raid late in 1862 on the little town of Marshall in Madison County, North Carolina, near the Tennessee border, to seize some salt, although the revolt was promptly and severely punished. Of the same general character was the rumor of a concerted attack on one of the North Carolina salt works by persons living in the surrounding counties, rendered desperate by their inability to get any of the commodity.[42] There were even rumors of plans to raid Saltville.

Scarcity was accompanied by its inevitable concomitant—soaring prices. In parts of the confederacy the highest priced article, relatively considered, of prime necessity was salt. For purposes of comparison it might be well to recall pre-war prices. On the eve of the conflict Liverpool salt sold in New Orleans at fifty cents a sack on board the vessel, one-fourth of a cent a pound, including the cost of the sack.[43] The first hint of an upward tendency which the writer has encountered came in August of the first year of the war when a newspaper insisted that a dollar a sack represented the full value of salt, a concession, it will be noted, above pre-war prices. Within little more than a month it could not be bought in Richmond for less than six dollars a sack, while it was selling in Raleigh, North Carolina, for eight dollars a bushel.[44] By January of the next year it commanded twenty-five dollars a sack in Savannah, Georgia. In Richmond in November, 1862, citizens secured it from the agent of the city council at fifty cents a pound, whereas private stores were selling it at auction at $1.30 a pound.[45] Governor Brown agreed that same fall to buy in behalf of the state of Georgia all the Louisiana rock salt a given firm could deliver in Atlanta at $7.50 a bushel or about fifteen cents a pound.[46] At the wells in Alabama the price ranged from $2.50 to $7 a bushel in gold, $30 to $40 in paper currency, though in the Trans-Mississippi region the average price seems to have been about $4 a bushel.[47] The state government of Virginia seemed able to drive a sharp bargain, for it contracted for 750,000 bushels at $2.25 in 1862 and for a fresh allotment of 40,000 bushels at $2 the next year, Virginia furnishing the sacks. At the same time the confederate government was obliged to pay $2.50 a bushel. In North Carolina it rose higher, $12 to $20 a bushel, and in the interior of Florida and Tennessee it commanded at times $1 a pound, a price often quoted by union officers.[48] Quotations for 1864 can be found as low as 30 cents in Georgia and at one point in Virginia; on the other hand, Governor Brown had to contract at $10 a bushel instead of $7.50 as before,[49] whereas a Mississippi agent agreed to $35 a bushel, a rate so high that the state repudiated the contract.[50] In general, however, the tendency was upward.[51]

At the very close of the war Governor Brown cited the price of salt in Georgia at five dollars a pound, while an officer gladly parted with that amount of money in Virginia for what might be termed a pinch of salt.[52] The domestic product, it might be noted, was of inferior quality compared with the Liverpool salt, and was often quoted at one-half the price of the imported article running the blockade.[53]

Naturally, efforts, which were only as successful as sumptuary laws usually are, were made to regulate the price of salt, along with that of other articles. The commanding officer in the Trans-Mississippi military district established a tariff of prices in June, 1862, for articles in common use, such as provisions, medicines, and leather. Salt was priced reasonably, considering its scarcity, at fifteen dollars per sack. Refusal to accept the price fixed constituted an offense for which the transgressor should be arrested and dealt with by headquarters "as such inhuman and disloyal conduct may deserve." [54] By special orders in the same military department a few months later, when the military took over the salt works in Arkansas and Indian Territory on government account, it was decreed that one-half of the amount manufactured should be sold to citizens at the low rate of $1.50 per bushel, or less if the cost of manufacture warranted, payable in kind and according to the schedule established in June.[55]

A schedule fixed for Texas in the fall of 1864 shows $2.50 per bushel of fifty pounds being paid for salt by the army authorities, a price which may perhaps be regarded as standard in that area for the army supply.[56]

The writer has not found evidence of fixing of prices for civilian consumption by confederate authority east of the Mississippi but action was taken by some of the states. Mississippi seems to have been the first to act. As early as 1861 the legislature, to curb extortion, levied heavy penalties for exceeding the scheduled prices; the state official schedule in April, 1864, fixed the price of salt at fifteen dollars a bushel.[57] Georgia's action, not so definite, left the question of extortion as one of fact to be ascertained by the jury in each case of indictment.[58] In Louisiana just before the loss of New Orleans in 1862 a tariff of prices for the more common articles of food, including salt, was established by the Moore government. A price of $2.50 per bushel was first made a sort of maximum in South Carolina, though later the restricted price was raised to ten dollars a bushel.[59] Florida seems to have contented herself with urging on her sister states the fixing of a maximum price.

As often happens when currency disappears or loses its value, society in the confederacy reverted to barter in its financial transactions, with salt a chief item in such exchange of commodities. It is an interesting fact that it had been a regular practice with the chief company at the Virginia salt works to pay for supplies in salt [60] so that that procedure during the war marked no departure at one point, at least, in the confederacy. Promptly in 1862, when the pressure for salt began to be keenly felt, proposals for an exchange in cotton, corn, and bacon were heard. The relative value of the two articles involved in the transfer enlists interest. A letter from a friend to Governor Pettus would indicate that in October, 1862, the exchange in bacon was being established in Mississippi at forty sides of bacon for one sack of salt.[61] Supplies, equipment, and labor were regularly paid for in kind at the various salt works. Salt was shipped from Alabama to pay for kettles and boilers; that state also fixed a regular nominal value of five dollars a bushel on the salt paid for labor.[62] Mississippi, finding it difficult to procure enough provisions for the hands engaged at her salt works in Alabama, decreed by law in 1863 that her general salt agent might exchange salt for provisions. The extent to which the confederacy reverted to barter may be understood when it is noted that the governments themselves shared in the practice. Governor Milton proposed that Florida tax the industry by appropriating one-tenth of the salt made, not only for the sake of the income but in order also to check the exorbitant price.[63]

The salt agent of the North Carolina works at Saltville worked out a peculiar, though just, technique in handling the question of barter. He refused to make exchanges for salt with individuals. He did make such contracts, however, with county commissioners in order to procure his supply of provisions, but the salt paid was counted as part of the counties' regular quota, so that counties obtained a preference in regard to salt only in point of time. He likewise exchanged salt for corn on the basis of bushel for bushel, computing the salt value at $1.50 a bushel, although citizens were offering him five bushels of the grain for one of salt. He contracted with the county commissioners of four pork-raising counties for 10,000 pounds of bacon, in consideration of which he allowed their counties ten pounds of salt per inhabitant by a given date, thus assuring those localities of a supply of salt to save their meat. He made a similar contract with the commissioners of Randolph County for 4,500 yards of osnaburgs at twenty cents a yard in order to clothe his hired negroes, and in order to make salt bags for counties which failed to send enough strong sacks. Another contract of like character was necessary with Lenoir County in order to secure 37,000 pounds of kettles.[64]

The central government at Richmond, appreciating the demand for salt and the possible depreciation of currency, arranged for large stores of salt which might be put to this very use of barter. The Virginia salt works were able to meet the demand for the army; but in view of the possible seizure of the Holston Valley by the federal forces, arrangements were wisely made by the commissary-general at the beginning of the war for procuring salt from various states of the confederacy, which supply would have permitted the use as barter of a large quantity derived from the Virginia works. The project was not, however, sustained by the secretary of war.[65]

The system of barter did not always work happily and produced complexities in connection with the private salt of the Virginia state superintendent at Saltville, as will be related later.[66]

Scarcity and consequent high prices brought into operation the law of substitutes, which meant, of course, inferior quality and the practice of extreme economy with the commodity. While passable, if not satisfactory, substitutes were promptly forthcoming for most of the common foods and beverages, such as coffee, tea, and sugar, salt ranked with certain medicines as practically without a substitute. It is true that we read occasionally of hickory ashes being mixed with salt. For lack of it, valuable food was sometimes utterly wasted. The Prussian officer, Heros von Borcke, for instance, tells of a windfall of a wagonload of oysters presented to his company, which were enjoyed with great relish for a few days, but which, offered without salt or pepper, soon became so nauseous that the mere sight of an oyster turned the men ill.[67] Green corn, a delicacy to the palates of starving men, tempting them to overindulgence, often proved doubly dangerous to their stomachs because of the lack of salt.

While salt in its manufactured form seems to the average mind to be very much of one quality, differing only as to grade of coarseness or fineness, it varied tremendously in its curative qualities. Mississippi, early recognizing that the various brines differed strikingly in degree of strength or weakness, set its geologist, Dr. E. W. Hilgard, to analyzing the relative values of the various kinds of salt available to her citizens. His results are interesting even to the modern reader and must have been noted with the keenest attention in 1862. His tests showed the Louisiana rock salt to be the purest of all; the salt made at Lake Bisteneau in northwestern Louisiana ranked second, surpassing the well-known Kanawha salt of West Virginia; the least pure was that from the wells in southern Alabama, which contained from three to five per cent of impurities. He pointed out that the latter might be purified by leaching it with water or brine and reboiling, a process which, obviously, would not reduce its cost. He invited submission of samples of brine from the various springs in the state, from which a feverish zeal was hoping to produce salt, and examined ten from various counties,[68] most of which he found altogether unfit for salt-boiling. It is an interesting fact that some southern chemists held that the presence of chloride of magnesium in salt was not a bad factor in the south, where, in order to preserve meat, the salt should dissolve as rapidly as possible.

Another interesting fact which developed as to the curative qualities of various salts was the point made by the state geologist of North Carolina in regard to the salt reduced from sea brine. Mr. Emmons pointed out to Governor Clark during the fall of 1861 that sea water contained ingredients detrimental to the best curative properties of salt. Simple boiling, he insisted, was not sufficient to free the brine of these bad ingredients, and so he proceeded to give information as to the best method of accomplishing that end.[69]

The planter world was agog in the fall of 1862 with rumors of substitutes available for curing bacon and beef, and it was for packing that substitutes counted, as it was for this purpose that the large amounts of salt were used. A newspaper in Alabama printed the statement that pyroligneous acid, made from any hard wood, would preserve meat as well as the best salt; that ten to fifteen gallons, which could be bought at fifty cents a gallon where wood was plentiful, would cure one thousand pounds of meat, and that the only objection was a smoky flavor imparted to the meat. The quantity of acid obtainable was nearly one-half the weight of wood.[70] Beef was cured with saltpetre and bacon with wood-ashes, often with unfortunate results. Some of those cures were so effective, one lugubrious sufferer relates, that in point of durability they "might have competed with petrifications themselves, and with fair prospect of success, supposing them to have been subjected to any agency of destruction short of Confederate hunger." [71] It is remarkable how the newspapers harked back to the practices of the Revolutionary War. A Revolutionary Tory of Albemarle had been refused salt to cure his pork, the *Daily Richmond Examiner* related in its issue of November 23, 1861, but his wife made good bacon with one peck of salt and an abundance of hickory ashes. "In applying the ashes," it directed, "it is well to have a bucket of molasses, and apply a portion with a white-washing brush to each joint. When well smeared, rub on the ashes, which will thus adhere firmly and make an impenetrable cement."

Every method which ingenuity could devise or imagine for curing meat without or with a minimum of salt was given a trial with the sad results that many experiments were failures and that the men were frequently served spoiled bacon. The commissary-general wrote with considerable scorn, "The high prices fixed by the county committees, and the fear that the commissioners of appraisement might not reach prices high enough to satisfy avarice, has doubtless stimulated every one who could spare any meat to bring it out, and the fear of its being fly-blown and spoiled in their hands has strengthened the patriotic desire of feeding the soldiers." [72]

The writer confesses to some mystification over the following sentence in Mrs. Varina Davis's book when it was first read: "Our salt had no preservative property." [73] It was obvious that it could not be dismissed with the explanation which disposes of the biblical paradox of salt which loses its savor, but knowledge of the inferior salt produced by hasty methods and of the practice of using the residue, known as "blocking", and of re-using the same salt many times over furnishes the explanation.

Blocking is a hard substance composed of lime, gypsum, and salt, a residuum which accumulates on the interior of the iron kettles, after their constant and prolonged use for evaporation and crystallization of the brine. If not removed at reasonable intervals, it breaks the vessels, as the iron and blocking expand unequally; furthermore, it is a non-conductor of heat and so is wasteful of fuel. It can be rendered sufficiently soft to be broken into fragments and removed by running pure, fresh water into the kettles, as was done at Saltville, through bored logs which ran along the furnaces. The fresh water absorbs and assimilates with the elements of the hard blocking, and by reboiling salt can be reduced from it, which is not as strong as that from the natural brine, but which the confederacy could not afford to scorn. It is clear from a report by Superintendent Clarkson that the Stuart, Buchanan Company up to the time of the seizure of their works by Virginia had been producing salt from this blocking. But, actuated probably by anger at the impressment of their works, they thwarted the effort of the superintendent to rent the blocking water furnace, which would have yielded forty to fifty bushels of salt a day, by threatening the new lessees with an injunction.

Evidently this process antedated the war, for Clarkson writes, "The lessors had long since erected (and the superintendent found it there when he took possession) immediately behind the two furnaces at the River Works, which adjoin each other, having a common smoke stack, a small furnace containing ten kettles, in which the blocking water and the fragments of blocking made at the River Works, were boiled and reduced to salt. Its cistern is furnished with and supplied by 'lead troughs' running along the sides of the furnaces into which the blocking water was poured, when dipped from the kettles, and by which it flowed off into this cistern."[74] The records do not show that he leased the blocking furnaces, but when Virginia preempted certain furnaces in March, 1864, she could clearly have exercised the same right toward these blocking furnaces.

Certainly, economy in regard to salt was learned and practised by large owners on the plantations to a degree which would have dazed and chilled their prodigal ancestors. A small supply of salt was hoarded and made to serve twice as long as would have been thought possible in ante-bellum days. Every grain was carefully shaken from the cured meat before the latter was used and religiously saved. Every successful experiment or new idea for producing or conserving salt was hastily and proudly passed on to neighbors and kin by a system of informal wireless. All the brine in the troughs and barrels where pork or beef had been salted was carefully dipped up, boiled down, and again converted into salt. Possibly the most dramatic piece of economy, and, certainly the one most often repeated, probably just because it was passed around by wireless and hence has lived in family tradition, was the following: the salty earth under the old smokehouses, impregnated with the drippings of years, was dug up and placed in hoppers resembling backwoods ashhoppers, made for leaching ashes in soap-making, with a trough underneath to catch the seeping water as it percolated through the hopper. The resulting brine was boiled down to the proper point, poured into vessels, and set in the sun in order to complete the rude process by evaporation. The residuum was unsightly in color, but it answered the purpose, especially for the stock, and was accepted without complaint, if not always gratefully.[75] Rice, grits, and hominy were washed or boiled in sea water and thus sufficiently flavored for the table. People were exhorted, if they must have salt food, to try corned beef, which called for less salt.[76] Naturally, a stock of salt was treasured and the grains guarded against the hazards of war as carefully as if they were indeed diamond crystals. One diary records how a barrel of salt, which had cost $200, was hidden in one of the slave gardens and the top covered with leached ashes to convey the idea that it was merely a leach-tub to secure lye for soap.[77]

The citizens' sense of economy prompted complaints in Virginia in regard to the inspection of salt required by law for the Virginia product, which was stigmatized as the mere guess of the inspector. It was also resented as a species of dishonesty that salt was bagged and shipped wet, a practice which had long been followed but which resulted in a loss of as much as twenty per cent in transit merely from Saltville to Richmond. On large quantities, such as 200,000 bushels, there was inflicted a heavy loss on the state and on individual purchasers. Not sufficient time for draining was allowed before the salt was thrown into the bin and weighed into the bags. It started, consequently, in an almost liquescent state.[78]

The desire for economy also prompted an effort to instruct the people scientifically as to the best method to cure meat with the modicum of salt. After salt began to be scarce, about December of the first year of the war, directions and proportions for packing meat began to be published. One recipe would, the paper declared, save hundreds of dollars in salt, as well as much time over the old process, but it is to be hoped that few planters trusted it at sight, as the statement is added that the planter vouching for it "would test it soon." [79] In 1863 Dr. F. P. Porcher was released from the confederate field service to prepare suggestions for the public on health and conservation. His directions on *Economy in the use of Salt* are worth quoting. "Coal consists mainly of the carbon in wood, which in burning forms a very drying heat. Most of our readers are familiar with the usual process of barbecuing large pieces of meat over coals. If such meat were too high above the coal-fire to roast, it would soon dry. When dry, a very little salt and smoking will keep it indefinitely. Like cured bacon, it should be packed in tight casks, and kept in a dry room.

"After one kills his hogs, if he is short of salt, let him get the water out of the meat by drying it over burning coals as soon as possible, first rubbing it with a little salt. Shade trees around a meat-house are injurious by creating dampness. Dry meat with a coal fire after it is smoked. You may dislike to have meat so dry as is suggested, but your own observation will tell you that the dryest hams generally keep the best. Certainly sweet, dry bacon is far better than moist, tainted bacon, and our aim is simply to show how meat may be cured and long kept with a trifle of salt, when war has rendered the latter scarce and expensive." [80]

CHAPTER IV

# EFFORTS TO PROVIDE A SUPPLY

## *By Individuals and by the Confederate Government*

AS WAS entirely natural, individuals and private companies were first in the field to set up salt works, for large bodies, such as legislatures, state or federal, get under way slowly.

The first effort to meet individual needs was, of course, directed toward the purchase of a supply. The constant search for salt, the appeals, especially to the governors as representing a public authority which should help its citizens, and the long distances traversed in the search for salt are indicative of the fact that the confederate planter tried to help himself. As the local supplies became exhausted, citizens of a given community would combine their funds and dispatch an agent in quest of salt for private use. For instance, a group of planters in Lafayette County in northern Mississippi sent an agent to the salt works in southern Alabama in the fall of 1862, where he contracted for the amount needed.[1] We also find Mississippians in large numbers making the long trip to Louisiana, two hundred and fifty miles and more, both to the salines in the northern part of the state, and, after its discovery, to the mine in the extreme southern part. We must remember, moreover, to measure the distance, not in terms of automobiles and macadamized roads, but in terms of mule-teams and boggy roads, poor at best, impassable in wet weather. It took one man four weeks to make the trip with a pair of mules from Rankin County in the central part of Mississippi to New Iberia.[2] Editors urged farmers individually to hitch up their teams and to set off for their own salt as the best plan of assuring themselves a supply. That this method was not neglected is clear, for we read of the roads in the vicinity of the noted salt works being crowded with wagons for great distances in all directions from the works, each awaiting its turn. Later in the war the danger of teams being impressed at the works by government became so great that only extreme necessity would induce planters to run the risk of losing them by driving to the various salt works.[3]

As is indicated elsewhere, thousands and thousands of citizens traveled hundreds of miles from the interior to the extensive seaboard to make salt for themselves and for their neighbors. While the coast was resorted to by residents of most of the southern states, the trip probably represented the greatest effort to the planters residing in the interior of Georgia, Alabama, and Mississippi.

The extortionate prices commanded by salt held out such promise of profit[4] that owners of even the pettiest spring with the weakest brine could scarcely resist setting up works. A list of places in Charleston, at which salt was made, was published in the summer of 1862.[5] Several samples of salt made from the water of the Ashley River were displayed in the office of the *Charleston Daily Courier*,[6] and specimens of a local product were placed in the window of the Tallahassee *Floridian*. Occasionally, the efforts to make the licks productive remind one strongly of the leaching of the smoke-house floors; one zealous farmer, for instance, in southern Mississippi was able to produce from two and one-half bushels of *earth* one gallon of salt. It is hence hardly surprising to read that the salt residuum was of a dark color, resembling brown sugar, the same as that of the earth from which the drippings came.[7] Such petty makers encountered the difficulties which beset the large producers, only in larger measure, such as the impossibility of procuring equipment—boiling vessels, barrels, and sacks. And such producers found that a fortune would not be readily amassed.[8] More wary owners of springs and lakes, who wished to escape the problems of manufacture, offered leases for the use of the water.[9]

Companies of varying strength appeared. One Mississippi company, the Strong, Cunningham Company, operating at Saltville, Virginia, secured an order for 100,000 bushels of salt from Governor Pettus for his state.[10] The reader cannot follow the records very far without becoming aware that the strong companies really producing salt in large quantities were relatively few: Stuart, Buchanan Company of Virginia, the Strong, Cunningham Company, the Georgia Planters' Salt Company, the Georgia Salt Manufacturing Company, the M. S. Temple Company, McClung and Jaques, and the Figh Company of Mississippi comprise the roll.

However, the companies seeking to exploit the states under the guise of patriotic devotion were legion. One instance may perhaps suffice. A certain prospective company wished to know what price Alabama would pay for the manufacture of salt in Florida, the state to advance the money and furnish the sacks.[11]

The suggestions emanating from individuals to various governors of possible sources of salt seem more disinterested. That from a man born in North Carolina but resident in Texas in March, 1863, when he wrote to Governor Vance on the subject, is certainly inspired solely by a desire to be of service. He pointed out an outcropping of a slate formation on his father's old farm, fifteen miles east of Charlotte, which was encrusted with a salty deposit sufficiently thick to be lifted on the point of a knife and the existence of a salt lick about three-quarters of a mile distant. He felt that the indications of salt warranted an investigation of the locality.[12] Professor Emmons, state geologist, who was consulted by Governor Vance, dismissed the suggestion lightly as a mere outcropping of rock found also elsewhere in the state. Again we find the state treasurer urging the state geologist of that same state to investigate a salt spring brought to the former's attention by the president of one of the railroads, a spot where salt had been made during the War of 1812.[13] The action also of Professor Kerr of Davidson College in bringing a beach near his salt works in South Carolina to the attention of Governor Vance is clearly inspired by no selfish considerations. He offered to secure a claim on the beach and then to hold it for the state of North Carolina.[14] After a thorough investigation he had concluded that the most favorable spot in the confederacy for the production of salt was the coast of South Carolina between the mouth of the Santee and Sullivan's Island near Charleston. He pointed out that it had the three qualifications of good transportation, abundant wood for fuel, and perfect security from the foe so long as Charleston stood. There were four salt beaches, which furnished a brine already so concentrated as to require less than half the wood necessary to make the same amount from ordinary sea water. He commented on the fact that during the War of 1812 salt had been made along that coast by the usual methods and also by the "cheaper European method of solar evaporation" and that some of those establishments had continued in operation many years after the war. He regarded it as feasible to manufacture there the entire supply needed by the confederacy by some of the cheaper methods of Europe at a very moderate cost per bushel, as the beaches were of indefinite extent along that coast.[15] It does not appear, however, that, aside from his own works, his suggestion bore fruit.

Such suggestions, pointing to other possible saline sources, were forthcoming in Mississippi as well as in the Carolinas. One such communication emphasized the possibility of "stone salt," indicating clearly from its date, late in 1862, the obsession which the discovery of the Louisiana mine had created.[16] Governor Pettus was lured by another individual to bore upon his land in Clarke County, Alabama, but the state agent, after several weeks of fruitless boring, advised abandonment of the work,[17] as he had encountered only fresh water.

A proposition from a Georgian, while not entirely free from self-interest, seems fair. He proposed that the state erect works equal in size to the works which he contemplated erecting on some land which he owned on the inside of Skidway Island, up the narrows where workers would be safe from attack; he offered the location and wood from his 500 acres of wood land gratis in return for the protection which he felt state works would afford his own, as he assumed that the former would be operated by a company of state troops.[18]

The confederate government set an example of zeal in discharging its duty in regard to providing a supply of salt for the army. More than once this subject claimed the attention of President Davis; congress passed legislation concerning it; and military officers could not escape the problem.

The actual legislation on the subject was slight, as was natural. In regard to the civilian population it was a matter for the individual state governments; in regard to the army it was the concern of the central administration, functioning under its war powers. On August 26, 1862, however, a bill was introduced into the senate, authorizing the president to expropriate and to work for the public use salt mines or springs. The Committee on Military Affairs, to which it was referred, reported a substitute, providing that the president could seize the usufruct of any salt mines or springs of private persons for just compensation, and appropriating $2,000,000 to carry the bill into effect.[19] On the same date, August 26, 1862, a similar resolution appeared in the house for the Committee on Military Affairs to study the power of the government to take control of the various salt works or to arrange with the owners for a regulation of the prices.[20] No legislation appears to have resulted.

On December 19, 1863, a Special Committee on the Manufacture of Salt was created by the lower house to report on the practicability of procuring a supply of salt in the vicinity of the salt wells in Virginia by mining and upon the comparative advantages of mining with other modes of supply.[21] The committee presented a most interesting report on February 15, 1864. The majority reported with considerable enthusiasm in favor of mining the salt. As an essential first step in the encouragement of this process, it found from the analysis of the Nitre and Mining Bureau that the natural product of the bed showed a chemical purity of 82 per cent. The great difficulties revealed in the process of manufacture by evaporation were the increasing scarcity of wood and the congestion of transportation over the Saltville branch of the railroad. The moving of the fuel consumed so much time of the road that it obstructed the carriage of the salt, even when there was a large supply of that article ready for distribution. The committee found that even under the boiling process large quantities of the salt water were being transported over the railroad to be evaporated at Glade Spring. It was pointed out that salt taken out in solid blocks could be transported without sacks or barrels upon platform cars, could be stored in large quantities for future emergencies, and would not be subject to easy destruction by scattering as was the case with salt crystallized from brine. The committee saw an added advantage in the fact that the rock salt could be shipped for purification to points conveniently located with reference to fuel and secure from interruption by the enemy, or it could pass directly to the consumers, who could themselves purify it in small quantities in their own wash kettles. Fuel, it was noted, was much more bulky than the rock salt, which might readily move toward coal deposits or forests to seek its fuel on its way to market. Transportation of the brine to great distances was, on the other hand, impracticable, since the salt constituted only about fourteen per cent of the water, whereas rock salt carried only eighteen per cent of excess bulk in the form of impurities. An added argument for mining the salt was the fact that it was readily soluble, much more so than blocking, and that the process of dissolving it with sea-water, as was done in England with the salt from the Cheshire mines, had proved the cheapest mode of manufacture possible. The additional cost of freight in carrying the impurities in the rock salt was more than counterbalanced by the loss incident to the existing system from imperfect drying, which resulted, as is elsewhere stated, in the loss of at least five pounds per bushel, and by the increased cost due to freight on fuel moving in to Saltville. Finally, a mine would be less vulnerable and more easily repaired than the machinery erected for artificial evaporation.

The legal difficulties with the owners and lessees of the wells in any scheme of mining by private operators were frankly recognized, as the lessees had secured rights only to the surface, while control of the surface by others debarred the owners from mining. But it was believed that power existed in the government under the impressment acts to take temporary possession and to operate a mine for the salt supply of the army. To settle the question definitely the committee suggested legislation to amplify the power.[22] One estimate put the possible cost of mining salt as low as one dollar a ton, equal to about thirty bushels of fifty pounds each.[23]

The telling arguments of the committee seem, however, to have fallen on deaf ears, or else the confederacy was too weak at that late date to undertake radical departures. Nothing seems to have come of the recommendation.

If the government fostered the salt industry, it must also be noted that she exacted a toll from it when it had become strong enough to warrant such an exaction. The tariff measure of May 21, 1861, laid a gentle duty of two cents a bushel on salt; but in 1863, when the shrinking revenues and increasing expenses of a desperate government necessitated seeking money from every possible source, salt was included with naval stores, liquors, agricultural lands, and agricultural products as commodities on which a property tax of eight per cent was levied, thus being classed with cash money as an object of convenient taxation.[24] Finally, it became subject in 1864, along with shoes and the more common food products, to a ten per cent profits tax.[25]

In order to insure a supply of salt for the country, the question of salt-making had to be considered in connection with the exemption law. Within five days after congress passed the first Conscription Act of April 16, 1862,[26] which declared every able-bodied white man between the ages of eighteen and thirty-five subject to military service, it began to work out its system of exemptions, but salt-manufacturers and their employees were not then included among the industrial classes which received exemption.[27] During the summer and fall the secretary of war, under the discretionary powers vested in him by the law, virtually extended the scope of exemption by enrolling and detailing men to work in government plants and for government contractors at the journeymen's price in order to meet the complaint of lack of labor to make army supplies. He justified his action by ruling that if the commodity being made were for the government, then the workers engaged in its production were in government service.[28] He stood firm, however, against Governor Shorter's request to direct the commandants of conscription to detail lessees and their employees to the state salt works in Alabama, though the order would have involved only about a half-dozen men.[29] Salt-making on government contract promptly became popular. Governor Milton pointed out to the Florida legislature in early September that many men of conscription age from other states, as well as from Florida, were very busy on the coast of that state, ostensibly making government salt, though he personally knew of ten men so associated in a company who had not made a bushel of salt in six weeks.[30]

Yielding to the clamor, Secretary Randolph recommended in a report to President Davis, dated August 12, that along with millers and tanners, who also coveted the privilege, salt workers "as essential to the prosecution of the war" be included in the exemption. Since he was supported by the chief executive, congress yielded to the pressure and included these three classes in the exemption in an amendatory act of early October.[31] All superintendents, managers, mechanics, and miners employed in the production of salt to the extent of twenty bushels per day were exempted from military service, this act being obviously intended not to apply to white salt-hands nor to the state salt agents.[32] It is noticeable that Governor Shorter carefully warned a local agent, when he sent him a certificate of office, that it would not protect him from conscription.[33] But mechanics and employees had to be enrolled and detailed to their special duties.[34] In order to guard against abuses,[35] exemptions were effective only for persons actually engaged in the industries enumerated, while in the execution of the act affidavits concerning the employment and the indispensability of the work were required.

By the fall of 1863 there was widespread dissatisfaction with the exemption system on the part of the people because of its abuses, and, on the part of the authorities, because it was operating so as seriously to undermine the entire conscription principle. Therefore, a new exemption act was passed February 17, 1864. It reduced greatly the number of exempted classes, while the president and secretary of war were given the power, within certain conditions, to exempt or detail men within these classes. Only men in a work of a professional or public service type were still exempted, while exemption of men engaged in agricultural and industrial production was left wholly under executive direction.[36] Salt workers were not mentioned by name but were included in the restriction laid on executive exemption: the governor might not authorize the exemption or detail of any contractor for furnishing supplies of any kind to the government unless the head or secretary of the government department making the contract certified that the personal services of the contractor were indispensable to the execution of the contract. Details should be from reserves forty-five to fifty years of age or from the regular army, when necessary, and the exemption ceased if the contractor did not apply himself diligently to his task.[37]

It was the intention of the Bureau of Conscription to draw all necessary labor, except that of agriculturists and mechanical experts, from the exempt classes: reserves, light-duty conscripts, and the invalid corps. Hence, there could be but few details from the able-bodied under fifty.[38] Although congress and the president were constantly struggling with the exemption question, agreed on the principle that the number of exempts needed to be reduced, they were utterly unable to agree as to the method of accomplishing it, and so the law of February, 1864, remained the law on the subject during the remainder of the life of the confederacy.[39]

The regulations governing exemption of state officers included also executives at the salt works. The first exemption act of April 21, 1862, had provided that all civil officers of the state, not required to serve in the militia, should be exempt from confederate military service. The amendment of May 1, 1863, included other state officers for whom the governor claimed exemption in order to secure proper administration of the government, provided such exemptions ceased at the end of the next regular session of the legislature unless that body exempted them by law. Finally, the act of February 17, 1864, provided for exemption of all state officers whom the governors certified as necessary for the proper administration of their respective governments.[40]

The governors had no power to exempt from confederate service, but they could recommend for exemption. While several of the states stood very staunchly by their governors, Alabama was conspicuous in indicating a certain suspicion that Governor Watts was abusing his privilege of recommendation and called upon him at the regular session of 1864 for a statement showing how much salt had been received from persons so exempted. His reply declared that "whenever it was shown that such contractors (for salt for state use) had in good faith commenced and were engaged in making salt under their contracts, I have recommended them to the Confederate authorities for exemption."[41] He was not able to give any figures as to the number recommended nor as to the amount of salt received from them.

While the governors rendered lip-service toward increasing the effective force of the army, their conduct was calculated to reduce it. Governor Brown of Georgia vigorously supported applications for exemption. When such a plea from Messrs. Seago, Palmer & Company, who were manufacturing salt for Georgia in Virginia, was denied by the secretary of war, appeal was carried to the chief executive. Brown pointed out with considerable force that to get the salt out efficient management was needed, that the work required men young enough to endure continued labor and fatigue, that men must be able to watch their investment if capital were to be attracted to the salt industry, suspension of which would entail suffering for soldiers' families, and he concluded by asking details for at least longer than thirty days.[42] After the passage of the law of February, 1864, he did not hesitate to recommend freely all persons engaged in the manufacture of salt for Georgia "as necessary agents of the state." It does not seem strange to think of the superintendent of the works, the president or manager, master of train, or overseers of the furnaces as officers of the state, but it is perhaps stretching even war conditions to think of the purchasing agent of forage, wagon master, or men furnishing wood to an engine as duly accredited state officials.[43] It is highly indicative of changes wrought by war that a governor wrote of one of the hands for Stuart, Buchanan Company in Virginia, "He is a very necessary man, and next to Musselwhite in importance to the keeping of the pumps in order, and also the lines of connecting logs to the furnaces, being deemed indispensable by the firm, to prevent delays in the water supply." In one letter Brown asked exemption for fifteen men for the Seago-Palmer Company, in another for ten for the Planters Salt Company, supplemented by a later request for exemption for four more employed by the same company.[44]

The numbers carried on the certified lists were large. In November, 1864, Governor Vance had 14,675 names on his list of North Carolina state "officers," only 26 per cent of whom, however, were credited to salt manufactories; Governor Brown of Georgia had 1,012, later corrected to 8,229; Virginia was reported to have 1,422; Alabama 1,223; South Carolina 233; Mississippi 110 (regarded as grossly inaccurate, for it was held to reach at least 4,000); Florida was credited with 109, East Tennessee with 39, and East Louisiana with 20. The report for the following February shows slight increases for all the states except Georgia, Mississippi (for which it remained the same), and North Carolina, but the decrease for the last-named state is very likely attributable to faulty reports.[45]

While the number of men exempted in some of the states was relatively high, it is conspicuous that the number detailed to salt manufactories is negligible, only 41 for Virginia, which had the largest number, and but 29 for Georgia, which ranked second, although at this confused period, February, 1865, it must be acknowledged that figures mean little.[46]

The exemption question presented peculiarly interesting aspects in North Carolina because of the presence there of a large number of Quakers and other pacifist groups. Although within the limits of the confederacy, they refused to join in the war. Furthermore, a jealous government could not forget that the Quakers, at least, had circulated abolition literature, for which some had suffered sharp punishment. Their attitude toward slavery made them, in the practical view, appear unfriendly to the new government, while the society served, in addition, as a refuge, even if perforce limited, for men seeking excuses from conscription. It offered but scant comfort that they just as steadily refused to fight for the union, though many were suspected of secretly favoring that side. Yet confederate government and conscientious objector could be fair toward each other as the following excerpt from a Quaker pen shows, "Considering how obnoxious their principles must have been to the Confederate government, it is to their credit that they often showed so much disposition to be lenient toward Friends." [47]

Under the conscription law of 1862 the Richmond government exempted adherents of the Friends, Dunkards, Nazarenes, and Menonites in regular membership from military service upon furnishing a substitute or upon payment of a tax of $500.[48] They proved no more willing to pay the tax than to fight, regarding it as a price for religious liberty.[49] By the amendment to the statute of June 7, 1864, the secretary of war was allowed to grant exemptions to members of the various denominations mentioned in the act of 1862 who were at the time of its passage members in good standing. There were some cases where Friends declined to pay the ransom and hence joined the ranks of deserters in the woods, caves, and dug-outs.

It would almost appear that the government of North Carolina viewed this sect as a source of revenue, for in May, 1862, at an adjourned session of the convention it enacted an ordinance exacting a hundred-dollar tax, or labor at the state salt works, or service as nurses at the state hospitals in lieu of military service.[50] However, a certain number must have preferred to render their service at the salt works rather than to defy the government, as the salt commissioner reported on September 19, 1862, that he had about 200 men who were exempt from military duty, one-third of whom were Quakers.[51] The governor, upon solicitation of those who felt much sympathy with the conscientious objectors, consented to allow those who were drafted to render their patriotic service to the state by sending a substitute to the state salt works. The salt commissioner further secured for them permission to pay eleven dollars a month as a commutation for service or for a substitute laborer, which money was to be used in making salt. Oddly enough, the Friends erected moral scruples against such labor as producing an article intended for the army and thereby protracting the war. They thus drew a distinction in an inconsistent manner between corn and salt in that regard, and failed to recognize that the salt was for the civilian population as well as for the fighting men. Members complying with the law were to be censured by the Society of Friends. This attitude aroused some popular resentment so that there was talk of seizing the Quakers who declined to comply with the law and of sending them to the hospitals as nurses, as the law directed.

At the very close of the war a special tenderness toward the salt-makers of North Carolina was manifested by Governor Vance, who had had a sharp conflict with the confederate military commander at Wilmington and with the Richmond authorities over the closing of the salt works on the North Carolina coast. After the salt works at Wilmington were broken up by General Whiting, the governor declared that salt workers home on furlough would be protected against confederate conscription, provided they joined the home guard and gave zealous service in catching the robbers and rogues who were then afflicting the country.[52] This stand was taken despite the fact that all holding furloughs had been ordered the preceding fall by the War Department to report to the camps of instruction.

The efforts of the central government to provide a supply of salt were confined, as has been stated, to meeting the needs of the army. They included all the usual devices: purchase, impressment of the commodity, and direct manufacture.[53] As a matter of course, the usual business-like procedure of placing large contracts with the more important manufacturers of the article was followed by the Subsistence Bureau of the War Department. Exerting a virtual monopoly over the product at Saltville, the bureau promptly in 1861 stipulated for all the salt it might choose to demand at seventy-five cents a bushel, but it found it impossible to obtain enough salt for its purposes. In order to encourage the enlargement of the salt works and to provide for the packing of meat, the department then made a contract with the proprietors for 10,000 bushels of salt per month for twelve months, with a clause authorizing an indefinite, intermediate demand, independent of the current monthly supply.[54] After the termination of the indefinite agreement on April 1, 1862, the government had a contract for 22,000 bushels a month for a year at the same price.[55] The needs of the central government became, naturally, the responsibility of Virginia when that state took over the works at Saltville. Hence, a contract was signed by the state on April 15, 1864, for 30,000 bushels a month at three dollars a bushel, deliverable at Saltville. The price, it should be noted, was less by one dollar than that at which the product was sold to her own citizens. To look after the government salt a man, living at Saltville, was commissioned commissary in charge of government salt at the works, and it was he who requisitioned the cars in 1864, and thus added to the friction which arose over transportation.

A casual reference in a Georgia document makes it clear that the confederate government had made a contract with the Planters' Company of that state, which was operating in Virginia.[56] When it proved impossible for the Virginia works to supply the needs of the central government, it made purchases elsewhere: 50,000 bushels were bought at Nashville at three dollars a bushel, as well as other amounts in West Tennessee in 1861; a commissary agent bought 500 sacks in Georgia;[57] and in December, 1863, a contract was made at Drake's Salines in North Louisiana with a J. C. Weeks for all the salt he could produce at the rate of ten dollars per bushel at the works, an engagement which he probably deeply regretted when his neighbors were selling at twelve to fifteen dollars a bushel.[58] The confederate government also purchased of a man named Stansberry, who was manufacturing the product at Tadpole Lake, in the Lake Bisteneau region.[59] Doubtless there were many other purchases of salt, of which the records have not been preserved.[60] All government orders enjoyed, of course, priority.

Impressment of supplies, which might technically be classified as enforced purchase, was in reality, in view of the worthlessness of the currency, seizure. Farmers and country merchants were required to turn over their stocks of salt.[61] State stores were preempted in the same way, and stocks entering the ports or being transported from one state into another were promptly seized by confederate officials. A rather amusing case in point was the seizure at Memphis in May, 1862, of some Liverpool salt belonging to Alabama. The state quarter-master then tried to have some confederate salt at Columbus, Mississippi, given the state in exchange, but apparently in vain, for in July we find him urging the governor to try to have some salt which had just entered a southern port from Havana and which had been seized by a confederate military officer turned over to replace the lot seized at Memphis. It is amusing to find the confederate officer at the same time graciously offering to sell Governor Shorter 300 sacks of this Havana salt at twenty dollars a sack, if it could be spared, and the governor declining it "on terms mentioned."[62] By September, however, the salt had been repaid to the state by some Virginia salt, for we find the governor carefully explaining to several persons that a certain allotment, which closely resembled Alabama salt, did not come from the state works in Clarke County but was Virginia salt turned over by the central government to replace the seized Liverpool salt.[63]

Such seizure by the confederate authorities seems to have led to a rather sharp, if brief, clash with Georgia. A confederate agent seized some salt in Griffin at the close of the year 1863, offering in payment thirty cents a pound. Dragged before the court by a promissory warrant, he was ordered by the judge to turn over the salt. Secure in his strength as an agent of the central government, he then defied the court order by again seizing the salt at the point of the bayonet. Appeals to Governor Brown from several citizens brought, characteristically, a sharp, firm telegram: "Let the Court pass an order directing the Sheriff to summon every man in the county if necessary, with all the arms they can command, and arrest the officer who has set aside the judgment of the court with the bayonet, and imprison him until he delivers up the property which the judgment directed him to deliver."[64] Unfortunately the records fail to satisfy our curiosity as to who won the victory in the end—the governor or confederate agent.

Finally, when confronted with the impossibility of purchasing an adequate supply, the government had recourse to direct action and itself turned manufacturer. Very early in 1862 a general was negotiating a three-year lease for some 4,000 acres of land in Perry County, Kentucky, with the privilege of using the owner's machinery to make salt and of using the timber and coal on the land. Although the government did not snatch at the lease, as the commissary of subsistence had already made arrangements for a supply from less precarious localities, it did ultimately sign the contract, for in December the very general who negotiated the agreement was in possession of these salt works.[65]

Again, the government acted promptly upon a hint from Governor Pettus of Mississippi that the newly-discovered rock-salt mine in Louisiana was not being worked to capacity,[66] and directed the special agent of the commissary department to whom the management of the mine was intrusted to open a shaft and to make arrangements for an extensive production, while the commanding officer, General Taylor, was instructed to give requisite protection and to promote the salt-operations. President Davis declared himself deeply conscious of the necessity for a vigorous prosecution of this work (in Louisiana), and anxious to secure to the country the full benefit to be derived from it.[67] The records are not perfectly clear, but it would appear that the confederate government assumed a certain supervisory direction through General Taylor. Furthermore, it sank two pits at government expense, which were known as confederate pits,[68] assembled many negroes, and organized a packing establishment at New Iberia to cure beef. During the succeeding months until April, 1863, large quantities of salt were transported by steamer to Vicksburg, to Port Hudson, and other points east of the Mississippi. Two companies of infantry and a section of artillery were posted on the island to preserve order among the workmen and to protect it against a sudden raid of the enemy.[69] This activity continued until April, 1863, when the works were attacked and destroyed by General Banks.[70]

The subsistence bureau was operating large works covering three-fourths of a square mile, at Saint Andrew's Bay on the gulf coast of Florida in 1863, and on the west arm of that same sheet of water in 1864.[71]

A few other evidences of activity by the confederate government can be produced, notably in northern Louisiana. It had a well at Lake Bisteneau, which area it claimed as public land, since it was an old lake bottom. The government also owned the land at Price's Licks, and hence asked no rent of private salt-makers on national property. In addition, it reserved some sites at Drake's Salines at Big and Little Licks.[72]

As has been indicated, in the Trans-Mississippi Department General Holmes in the fall of 1862 ordered all salt works in Arkansas and the Indian Territory which were not producing to their greatest capacity to be taken over under fair compensation by a government agent.

That officer was vested with authority to impress, if necessary, the requisite labor and material, one-half of the product to be set aside for army use, the remainder to be sold to the civilians on the basis of $1.50 a bushel or less, if justified by the cost of manufacture. It was furthermore provided that the salt might be exchanged for articles of subsistence. Each citizen, however, was limited to a reasonable supply for domestic consumption. This order was inspired by the inconvenience caused by the insufficient output of salt, by the extortionate price, and by the threatened lack of meat provisions for the succeeding winter. The promise was held out that as soon as the emergency was past the order would be rescinded, but the writer has encountered no evidence of such revocation.[73] The following summer it was recognized that the entire Trans-Mississippi must be self-sustaining in regard to salt as well as shoes, clothing, powder, lead, ammunition, and wagons.[74]

Finally, army officers were frequently forced to set a group of soldiers to work at some neighboring salt works or lick to furnish the company with a salt supply. While most of the specific instances occurred in the country west of the Mississippi, in 1861 a general sent a detachment to Brashearsville on the north fork of the Kentucky River not only to supply his current demand but also to enable him to cure his future meat ration.[75]

One of the most interesting bits of governmental activity on this subject which has survived is the effort to promote the production of salt by solar evaporation, a project linked up with one of the most fascinating characters encountered in this study. It has already been indicated that a French geologist, M. J. Raymond Thomassy, had predicted the importance with which a war would invest this simple commodity.[76] This interesting person had made four or five tours of inspection of the Mississippi delta, one just on the eve of the war, resulting in his *Géologie pratique de la Louisiane*. His deep interest in the south and his knowledge of her ability to supply some of her needs from her own resources, but especially a very peculiar interest in salt and the method of making it by solar evaporation, evidently prompted him immediately upon the outbreak of hostilities, late in 1861, to press his views upon the legislatures, first of one state and then of another, upon editors,[77] and finally upon the government leaders in Richmond. He finally convinced the secretary of war of the value of his views.

The method of making salt by solar evaporation, which he so ardently advocated, had already been in practice in southern France and in Italy for many years. Naturally, the concentration of the crystals by natural evaporation is much cheaper than by artificial heat. The cost to the manufacturer of making salt in France from 1835-1855 was about two cents a bushel. By some new improvements introduced by M. Thomassy into the Italian salt works in 1848, a bushel was produced from the weak brine of the Adriatic Sea for one and a half cents.[78]

He had already in 1859 addressed a memorial to the several southern legislatures upon the subject of the new process invented by himself, urging it upon them with strong arguments. He pointed out how southern salt made from sea-water would prevent a drain of nearly $2,000,000 abroad; how its trifling cost would result in its application by millions of bushels to industry and to agriculture; how its bulky character made it an excellent article with which to load the freight trains as they discharged their agricultural commodities at the coast; how the superiority of the sun-dried salt would quickly displace the deliquescent Liverpool salt in the north, especially for the fisheries; and, finally, how it would increase the shipping of the south.[79] The first state which he seems to have approached, at least the first in which he received a friendly welcome, was South Carolina. In mid-November of the very first year of the war, an enterprising house at Charleston was reported as having engaged the services of Professor Thomassy and as erecting salt works on a grand scale. It would appear that Thomassy issued a prospectus, proposing the organization of a joint-stock company for manufacture of salt from sea-water, expressing confidence in the possibility of erection of a manufactory for $75,000, capable of producing 100,000 bushels of salt the first year at a cost of only ten cents a bushel, with possibility of expansion to 400,000 bushels at a cost of only six cents per bushel.[80]

The second state from which he seems to have received a favorable hearing was Alabama, where in December, 1861, at the regular annual session the legislature authorized the incorporation of four men, R. Thomassy heading the list, with a capital stock of $1,000,000 to purchase such real estate upon the rivers, waters, bays, and sea or gulf shore of that state, as might be necessary for the manufacture of salt, and with power to construct inland passage-ways, canals, or such other means of transportation as might be necessary. This was a purely private and commercial venture, for the corporation was vested with the right to establish depots and agencies for the sale and distribution of salt in the populated districts of the state. Alabama seems to have been sincerely interested in encouraging the production of salt within the confederacy, for one section of the act of incorporation expressly empowered the company to accept such privileges as might be granted it by any of her sister states.[81]

Possibly the Frenchman found that he could not work happily with Alabama business men. Whether for this or for other reasons, we find his grant from the Louisiana legislature just about a year later entirely personal. He was granted the exclusive right for twenty-five years to use any salt springs or saline waters made available by him, together with the right to use the materials on the public lands of the state for the purpose of making salt, either by artificial or atmospheric evaporation. Any saline waters which had already been used for salt-production were excluded; he was required to begin the making of salt within three months after the discovery of any salt springs; Lake Bisteneau was expressly excluded from the grant; and his rights were contingent on his beginning salt-manufacture within six months. The grant carried with it the use of timber or other materials found on any section of land yielding brine in case such were necessary to enable him to carry on his salt production on the largest scale.[82]

Since the central government claimed the Lake Bisteneau region as public land, and hence even a foreigner could repair there with confederate authorization, there is reason to suppose that the Frenchman, named Thomassy, who appeared at Tadpole Slough in the Lake Bisteneau area the latter part of 1862 or in the spring of 1863, was this very distinguished geologist. He leveled off a large area on the slough, dug a large well, and laid off the ground for a series of basins in which to make salt by his pet process. He began the erection of a salt house on the eastern end of what is known, probably in his honor, as *Frenchman Island*. As he was a man of fastidious tastes, he employed a man to accompany him to cut away all the roots and to fill the holes on the road leading to the works in order to save himself rough jolting in his buggy. These parlor methods were regarded by the local people with such distinct disfavor that they scorned his whole scheme as visionary. When he had the leveled ground laid out with stakes ready for excavation, a party of rough Arkansan "hill-billies" arrived on the scene. Concluding that Thomassy had too much ground for one man and probably viewing the particular locality which he had leveled with envious eyes, they immediately began to dig wells there and replied to Thomassy's protests in the rough vernacular of the time and region by telling him that, "if he didn't dry up and leave the country they would put him up a tree at the end of a rope." In a period of war, victory went to might with the result that Thomassy departed, declaring that he would carry his complaints to President Davis.[83] Whether or not he carried his grievance to the chief executive does not appear, but that he pressed his project, which seems to have been almost an obsession with him, upon the high officials at Richmond, and with success, is certain.

The salt mine which had been recently discovered at New Iberia could not escape the eagle attention of such an enthusiast on the subject of salt as M. Thomassy. He had visited the salt spring on Judge Avery's plantation on Petite Anse in 1857 and instantly traced the source to masses of rock salt beneath the surface. He made a second visit to the island shortly after the discovery of the rock-mass in order to make a more thorough examination of it. His report of this visit brought the discovery before the scientific world.[84]

Sometime during this period it would appear that Thomassy directed his attention again to the salines in southern Alabama, for his next venture links up the Richmond government with this location for his project. It is certain that he had at some time before late December, 1862, visited these wells,[85] and it is not unreasonable to suppose that it may have been after he was driven from the Louisiana salines when he was on his way to Richmond to complain to President Davis. He would undoubtedly know of the Alabama salines and his fertile mind readily conceived of utilizing this area.

A law passed by the Alabama legislature in December, 1862, shows that the confederate government wished to negotiate a lease for the purpose of making salt by solar evaporation at the upper saline reserves of that state in Clarke County. The governor was authorized to make such a lease for ten years for an area of not more than twenty acres under the express condition that such activity should not injure the operations of the state or of its citizens at the reserves.[86] A few days after its ratification, we find Secretary Seddon asking Governor Shorter if the state could not allow one hundred instead of twenty acres,[87] since the method called for large acreage for the many shallow pans required. Unfortunately, the legislature had already adjourned, but the governor, sympathetic to the plan of solar evaporation, assured the secretary of war and the commissary-general that he would not suffer the enterprise to fail for want of area for evaporating fields. He declared that the works might proceed with the strong confidence that the legislature would confirm whatever he and Professor Thomassy might do. He felt sure that if the legislators had been fully advised as to the extent and importance of the contemplated works, they would have drawn the bill so as to meet the views of the government.[88] But whether Thomassy found the Alabama brine too weak, or what the cause, there is no evidence of confederate works in Clarke County. The writer is forced to the conclusion that the project did not materialize.[89]

Yet one more reference to Thomassy, and he passes from the records and from the confederate stage. On January 5, 1863, the Chief of the Engineer Bureau issued the following order to a subordinate at Atlanta: "As soon as you can get the bridges under construction, you will meet Professor Thomassy at Atlanta, to receive his views in detail, with drawings of the works necessary to the production of salt by solar evaporation. The professor then will go to Europe, by authority of the War Department, to employ persons skilled in the manufacture of salt.

"You are authorized to employ an assistant engineer to perform the local duties in preparation for the salt works." [90] It is possible that these local works, judging from the synchronism of dates, refer to Alabama and not to Georgia. But Thomassy died in 1863 so that the confederacy perished without ever seeing the consummation of the long and tenaciously cherished plans of this French friend.

CHAPTER V

# STATE EFFORTS TO PROVIDE SALT

## *Embargo and State Aid*

WHILE EFFORTS were being made by the confederate authorities to provide the necessary salt for the army, a task which the various states were obliged to share, the responsibility of providing the civilian population with that article rested squarely on the state governments alone. Under the impetus of recommendations from the governors and of their own initiative, the various state legislatures devoted much time and thought to the solution of the salt problem. The subject occupies much space in the governor's annual messages; numerous laws were spread on the statute-books; special committees pondered on the subject. At least one legislative body was convoked in special session because of its urgency—that of Mississippi in December, 1862, to which Governor Pettus declared salt the "Most pressing want at the present time." [1] Probably no committee had more serious or difficult problems to untangle than the Joint Committee on Salt in Virginia, which functioned in 1861-62 and was revived in 1863.

A search for new salt deposits was straightway encouraged. Governor Brown, upon the urgent suggestion of several citizens of Georgia, offered a reward for the discovery of salines, "as it is most probable that we cannot protect salt-works upon our coast against the gunboats of the enemy without a large force." He offered a reward of $5,000 for discovery of a salt spring, ten miles inland, which could produce no less than 300 bushels of salt a day. The reward was never, so far as the writer has been able to learn, claimed.[2] At about the same time, the spring of 1862, the state geologist of Mississippi, E. W. Hilgard, was directed by Governor Pettus to report on the salt resources of that state. He visited a number of reputed salt springs and licks in Raeburn and Hinds Counties, confirming his opinion from previous field work that no natural brines of sufficient strength to justify exploitation existed. The licks, not at all infrequent in that section, were merely superficial accumulations, resulting from the evaporation of weak, and mostly impure, brines from the surface of a porous soil.[3] Through a circular that useful gentleman distributed general directions for making salt at these salt licks by leaching the top earth, wherever indications justified the attempt. At the same time he invited the submission of specimens of salt or earth for his examination. He received, in consequence, as has been told earlier, numerous specimens, and found that some samples, from Rankin, Hinds, and Pamola Counties yielded salt of superior quality, but that none seem to have proved of important market value.[4]

The idea of boring for salt water was popular in North Carolina and Virginia. Authority was vested in May, 1862, by the legislature of the former state in its salt commissioner to make such borings wherever he deemed it advisable. The investigations of the state geologist, Professor Emmons, encouraged Governor Clark to hope for results in Chatham County and to contract for a boring machine. Whether Worth, the salt commissioner, was cool to the idea, or whatever the cause, the governor's enthusiasm does not seem to have brought action.[5] After the loss of the works on the coast, the legislature returned to the subject in December, 1864, instructing the salt commissioner to examine the salt marsh in Bladen County for springs, and to sink shafts at his discretion.[6] Some legislators urged the Virginia assembly to direct inquiry into the expediency of experimental borings for salt water near Clifton Forge on the line between Rockbridge and Botetourt Counties, or near Big Lick in Roanoke; another member viewed with friendly eyes, Lick River Ford in Botetourt County.[7]

The first reaction to the situation, as state officials began to be aware of the scarcity of salt, was an embargo on its exportation from the state and prohibition of monopoly with a view to extortion, accompanied usually with provisions for seizure of the article by state officials. The two actions were so closely linked that it is almost impossible to handle them separately. Florida was first in the field with such a prohibition, laid by legislative action in the November session of 1861. Salt was enumerated in a long list of provisions which were denied exportation except for the use of the state—presumably for soldiers of the Florida line—or for the use of the Confederate States. The same act forbade purchase of provisions of any kind for speculation or for sale at a price which yielded a profit of more than thirty-three per cent.[8] This same omnibus bill authorized the governor to seize beef, hogs, salt, or provisions for the sustenance of the army with the condition that there be just compensation.[9] That some seizures of salt occurred is clear, for we have the record of one merchant of Apalachicola, whose salt was seized and sold at five dollars a sack.[10]

Governor Brown of Georgia acted with conspicuous promptitude. By executive initiative on November 20, 1861, he ordered for the use of state troops the seizure by the Commissary Department of any salt found in large quantities at Macon, Columbus, and Atlanta, in case a price of more than five dollars a sack were demanded, or in case it were withheld from sale for purposes of speculation. The seizing officer, however, must tender five dollars a sack in payment.[11] Orders to the superintendents of the main railroads of the state directed them to detain all salt unless shipped under affidavit that it was for personal use. The governor justified his action under the constitutional clause which provided for seizure of private property for public use with just compensation.[12] He directed the mayor of Augusta on November 21 to seize the salt which was at the station, and to hold it subject to the governor's order. A few days later he also required him to take in custody from salt dealers some 2,000 sacks, which he was charged to keep within the state. The official replied that in the excited state of feeling among the merchants, his authority would not be recognized, and suggested that the governor send a special agent to execute the orders.[13] The refusal of the superintendent of the Central Railroad to ship salt from Savannah brought a similar clamor in that city several weeks later.[14]

The governor was not fully sustained by the assembly in his views with regard to extortion, for, instead of passing the bill desired by the executive to make it penal to sell salt and specified articles at more than sixty per cent advance on the prices of April, 1861, the question of extortion was left as one of fact to be ascertained by the jury in each case of indictment. This fact necessitated a change in Governor Brown's orders from payment of five dollars per sack to an amount ten per cent above the cost price sworn to by the owner, to which he could add the cost of transportation.[15] It was a source of regret to Governor Brown that the class of extortioners who had large lots of salt locked up in city cellars had not been reached.[16] A few months later he suggested that citizens do all they could to "permit speculators, who have a supply on hand for the accommodation of the people at fifteen to twenty dollars per bushel, to hold it till the end of the war, when they can probably afford to sell it much cheaper." [17]

One of the early acts of the Alabama legislature at its regular annual session in November, 1861, was a law empowering the governor to seize for public use under just compensation all salt stored or held for high prices, and making the secreting of it illegal.[18] That seizures were made by Governor Moore we know, for an appeal from several citizens of Eufaula, dated November 11, pled for the release of some salt because the owners had acquired it normally, had held it openly for sale, and had repeatedly refused to sell to speculators at prices higher than they were receiving at retail. He seized 1,400 sacks from a warehouse on the Montgomery and West Point Railroad, where circumstances warranted the suspicion that the owners intended to violate the embargo.[19]

The convention which met in Virginia in 1861 ruled that the governor, whenever in his opinion the exigencies of the public service required it, might prohibit the export of provisions of every character.[20] In October of the next year Governor Letcher sought the opinion of the attorney-general as to his power to impress salt which might be in the hands of speculators. The latter official held that the executive had clear power to seize salt from any person, where he deemed it necessary to secure a supply for the people of the state.[21] A few days earlier he had issued a proclamation forbidding the exportation of salt from the state, in accordance with the law of October first preceding, except in cases of contract with the confederate government.[22]

Though North Carolina moved swiftly, as we shall see, in positive measures to furnish a supply of salt by her own manufacture of that commodity, her legislation on the negative side to prohibit loss of what she already had was slower.[23] It was not until about mid-November, 1862, that Governor Vance took the first steps toward an embargo in order to call a halt to the "wicked system of speculation" by requesting the transportation companies to carry no more salt out of the state. Notices signed by the agents of the various railroads, notifying the public of the new rule, promptly appeared on the street corners. This action was followed by a formal proclamation by the governor, issued November 17, laying an embargo for thirty days on a number of articles of clothing and food, salt heading the list. Such action did not, however, prohibit the exportation of salt purchased prior to the date of the proclamation, as he later explained, nor transportation of salt for individual use, nor for charitable distribution by counties or towns, nor the entry of cargoes into North Carolina ports from abroad.[24] Despite the host of appeals which instantly began to pour in upon him, indicating numerous exceptions which would be necessary, he pressed for legislation on the subject in his annual message.[25] Accepting his recommendation, the general assembly immediately laid such an embargo on any article of prime necessity, except those held by agents of the confederate states or of the individual states, to be effective for thirty days—hence to December 22, 1862.[26] Repeated renewals kept it effective, however, at least until after May, 1863.[27]

Naturally, all sorts of deviation from the rule had to be allowed, aside from those covered by contracts made prior to the promulgation of the proclamation. Certain South Carolinians, making salt on their own coast, were forced by circumstances to deliver it for shipment at a North Carolina point and were caught with large lots of salt awaiting railroad transportation at that point.[28] One philanthropic man from Columbia, South Carolina, could scarcely be denied the privilege of exporting an allotment of 300 bushels, purchased at Wilmington, which he proposed to sell at cost to his poor fellow citizens, especially after he had appeared in Wilmington with supplies for the sick there during the yellow fever.[29] The military had to be satisfied, as usual, that a *bona fide* contract had been signed previous to the governor's order.[30] One Virginian, in ignorance of the new ruling, ordered salt from Wilmington for a distiller in order that the latter might cure his pork for the government.[31] Charges occur that the railroads disregarded the rule, shipping everything offered; and in April, 1863, the salt-makers got up a petition to ask Governor Vance to withdraw the restrictions on salt, on the ground that the price was declining in consequence, so as to discourage private makers. The inevitable ultimate result would be to increase the price.[32]

The law to prevent monopolies, extortion, and speculation in articles of general use appeared in North Carolina in the session of 1862-63, though under relatively slight penalties—a fine of $500 and imprisonment for six months. Concern to protect the existing stores of salt prompted a resolution that the governor inquire into the expediency of removing the salt, both state and private, from near Wilmington to some place in the interior more secure from the reach of the enemy.[33]

In South Carolina Governor Pickens, acting at his own discretion, seized more than three hundred sacks of Liverpool salt in the different towns of the state in the fall of 1861 to be held and sold in small amounts for the relief of soldiers' families, as a measure against monopoly and extortion.[34] A plan with unique features, by which it was hoped that extortion and exportation could be prevented, was devised in this state and enacted into law in February, 1863. But the hopes were not justified, for a spirit of speculation soon manifested itself which brought a virtual monopoly of such articles of prime necessity as flour, corn, bacon, and salt. They were being in part withheld from the market, and in part exported out of the state. In this situation Governor Bonham interposed, availing himself of his constitutional power, to prohibit the export of provisions for a period of thirty days. At the assembling of the legislature at an extra session in April, 1863, he warned members that legislation was necessary to continue the prohibition and to carry it more fully into effect. He also recommended some legislation to arrest the purchase and monopoly of articles of prime necessity, even when it was not intended to export them beyond the state.[35]

Louisiana, still wholly under the control of the confederacy during the first year of the war, acted promptly to preserve the valuable salines in the northern part of the state for the benefit of all her citizens and not alone for the benefit of the greedy few who might try to monopolize these sources of wealth. She withdrew from sale her public lands on which were located the salt springs with the exception of the public lands located on the Gulf coast.[36]

Encouragement and aid to individuals and to private companies was a third method, as simple as it was obvious, which could not fail to present itself to the minds of legislators and executives. Governors lent aid and information in every way they could. Governor Pettus took comfort in the reflection that in consequence of the information and assistance he had furnished, many individuals had succeeded in supplying themselves and their neighbors with salt.[37]

Manufacture of salt on the coast was stimulated by various devices in almost all the states touching the Atlantic or the Gulf coast. In Georgia the Joint Committee on Salt Supply, concurring with the report of the Committee on Transportation, requested the governor to encourage salt-making on the coast by directing the state railroad to afford special facilities for transportation of salt, so far as possible, by sending special trains or cars or by giving reasonable preference to shipments of salt and of materials needed in its manufacture, and by lending assistance in procuring iron and kettles for its manufacture.[38] This joint committee reached the conclusion that coast operations, though involving some risk, were practicable. While it felt that large operations might be difficult at any one point, experience had proved that two or more counties, or even individuals, could cooperate successfully for their home supply.[39] Governor Brown recommended a certain citizen, searching the seaboard for a suitable site, to the protection of the military authorities, and expressed the hope in July, 1862, that every citizen living near the coast would make at least his own supply.[40]

The legislature of Georgia faced the scarcity in December of the first year of the war by appropriating $50,000 to be advanced without interest to any companies engaged in salt-manufacture, including those already established, the sum loaned to be duly secured by mortgage and to be refundable after the war.[41] Governor Brown, with his usual determination, tried to use this loan fund to assure the people of a supply of salt, if one may judge from one contract with a Dr. F. J. Clarke, whereby the executive allowed an advance of $5,000, provided he might control all the good salt produced until he had claimed $15,000 worth at four dollars a sack. Furthermore, he agreed, if satisfied of the success of the works, to advance an additional $5,000 under the condition that he could command for consumers all the salt made for twelve months at four dollars per sack.[42] Another advance to Stotesberry and Humphries of $10,000 was made by the state, but failure on their part to take active steps toward manufacture led the governor to recommend that they be required to proceed at once or to refund the money.[43] The following session saw a much larger appropriation for salt, $500,000, in the use of which the governor was allowed a wider discretion, since he might use it to purchase or to make salt, as well as advance it upon proper security to individuals or associations for the purchase, manufacture, or transportation of salt. He was allowed to use $50,000 of the sum to send trains to Saltville, Virginia, and elsewhere to aid in transporting salt to Georgia, provided it was not intended for speculation.[44]

Two Georgia companies, the Planters Salt Manufacturing Company and the Georgia Salt Manufacturing Company, secured contracts for brine at Saltville and began to operate, under the special sanction of the state, with the distinct understanding that they were to sell to citizens at prices which covered only costs. Stuart, Buchanan and Company were under contract to furnish from their wells enough salt water to the Planters Salt Company to keep 140 kettles of 120 gallon capacity at regular work to their full capacity for the period of the war.[45]

Governor Clark urged North Carolinians, especially residents of Washington County, to engage in making salt.[46] An ordinance, passed by the convention of 1861 in order to promote salt manufacture in the interior of the state, allowed to a company operating in Chatham County exemption from military duty for the president and six operatives for six months, except in case of invasion or confederate requisition, and also exemption of the capital stock from taxation for six months.[47] A report of a special Joint Committee to the assembly of 1862-63, dismissing the idea of state subscriptions to stock or of a bounty to private companies, because of accessibility to the coast, recommended to the eastern counties domestic manufacture as practised in the War of 1812, since large shallow salt pans of cast iron could be had at a cost within the reach of anyone.[48] The state recurred to the idea of encouragement of private salt manufacture in the interior after the works at Wilmington had been broken up, for at the session of February, 1865, the governor was authorized to sell or rent any portion of the machinery to any salt producer located in the interior.[49]

The state reservations in Alabama were thrown open to all citizens of the state by circular letter of the executive issued June 30, 1862; they were invited and urged to make salt there for their own family consumption. All were forbidden to sell salt on the premises without first securing a lease,[50] a regulation clandestinely violated. When it was apparent that the state works could not get under way in time to provide a large supply for the packing season of 1862-63, the salt commissioner added his plea to that of the governor that the people help themselves by making their own salt at the wells in Clarke County or on the Gulf coast.[51]

Meanwhile other action to encourage production of salt had been taken. Governor Moore at the called session of the Alabama legislature in October, 1861, contented himself with recommending the lease of the Clarke County salt springs to some one who would obligate himself to commence the manufacture of salt at once.[52] The legislature, however, in order to encourage manufacture of salt in the state, responded to his suggestion with an act, which authorized the lease of all public salt springs in Clarke County to individuals and companies for terms of ten years, and offered a bonus of ten cents per bushel if the salt were sold to citizens of Alabama in small quantities, the bonus being limited to 500 bushels in one year. The lessee was to have the privilege of getting wood and other necessary materials from state lands, but must give a bond of $5,000 to commence the work in not less than three months from the date of his lease (later amended to two months), and was restricted in price to seventy-five cents, cost of transportation to be added if the salt were sold elsewhere than at the point of production.[53] In order to encourage the manufacture of salt "at the earliest day practicable," an amendatory act, passed a few days later, authorized the advance of $10,000 from the state treasury to any responsible person or firm, for the purchase of materials required to commence operations, such as machines, boilers, and other indispensable fixtures. The governor was to retain out of the bounty fund five cents per bushel until the amount advanced was returned. If the war terminated before the state had been whollv reimbursed, the governor was to relinquish the unpaid portion but to take over all improvements on the state lands.[54]

The low price at which the lessees were required to sell under their contracts naturally drew to their works in numbers far beyond their ability to supply purchasers, who became an absolute hindrance to the work in these remote and isolated locations.

A second amendatory act passed about two weeks later increased the maximum allowable price of salt to $1.15,[55] while also authorizing the division of the $10,000 loan between two persons or firms. It was, however, made the duty of the governor to secure the efficient working of the salt springs and to reserve to any citizen of Alabama the right to make salt for his own use at either of the state reservations, provided the operations of the leases were not obstructed.[56]

In conformity with the act, the lower reservation was leased to John P. Figh and Company, to which firm was advanced $6,000 to aid in reconstructing their furnaces and in other necessary works. Later, to increase their output, Governor Shorter proposed to them an additional advance of $4,000, on condition that they produce after April, 1862, one hundred bushels of salt per day, the state to be allowed to purchase all the surplus after prior contracts and payment for provisions and incidental expenses had been discharged by the sale of salt.[57]

The legislature at its next session radically altered the contract with the Figh Company. The company was obligated to deliver to a state agent two-fifths of all the salt manufactured, the state abandoning the bounty but in return leaving the price unlimited. The lessee was assured exclusive use of all timber on his lease.[58]

South Carolina, at this same critical time, December, 1861, authorized the formation of two joint stock companies for the manufacture of salt, the state agreeing to subscribe $5,000 as soon as the company had commanded $5,000 in $100 shares, with modest first installments paid, though the state naturally disclaimed any responsibility for company obligations and limited the company to twenty per cent dividends.[59]

In Arkansas an appropriation of $300,000 was made in 1862 to encourage the manufacture of salt, iron, and cotton cards, sums to be allocated under bond at the discretion of the governor, repayable with six per cent interest after an interval varying from two to six years. Persons who erected machinery on public lands should receive a deed for lands up to 160 acres, with the condition of reversion of the property to the state in the event of failure.[60]

## CHAPTER VI

# STATE ACTION

### *The States as Salt Manufacturers and Dealers*

THE FOURTH method by which the states sought to provide a supply of salt for their citizens might be termed direct action, as the states themselves went into business, entering the market as purchasers and manufacturers of salt. These two processes are so intermingled and were resorted to so interchangeably that any effort to treat them separately would lead only to hopeless confusion. State agents were commissioned to secure salt in either or both ways. This method constitutes by far the most important field of the state activity.

Governor Moore of Alabama without legislative warrant or special appropriation boldly devoted a portion of the military fund, available for the quarter-master and commissary departments by an act of February 5, 1861, to the purchase of a quantity of salt for distribution and sale to the people of the state at reasonable rates.[1]

Realizing that the yield from the salt region of Clarke County would fall far short of the demands of even the southern and central portions of Alabama, and that its transportation to the northern counties would be attended with delays and waste, Governor Shorter obtained in 1862 from the proprietors of the Virginia works a contract for Alabama to manufacture at the Virginia wells, from which source he hoped to secure a supply for the northern part of his state.[2] The difficulties in the way of operating state works in Virginia led to a transfer of the contract to McClung and Jaques of Knoxville, Tennessee, who undertook to produce salt for Alabama at $1.75 per bushel as long as the governor thought necessary. On the one hand, the state obligated itself to take 100,000 bushels in order to assure the company against loss; on the other hand, the company must continue to deliver after the 100,000 bushels had been furnished, if Alabama demanded it, as the right to terminate the contract inhered in the state. The state further contracted to supply the sacks.[3]

The difficulties which beset them so delayed McClung and Jaques that they were not under way before October, despite the urging and insistence of Governor Shorter.[4] In the spring of 1863, this firm notified the governor that on account of an increase in the price of wood from five to thirty dollars a cord, they could not manufacture at the price agreed upon. Fearing loss of the salt for northern Alabama, the governor made a new contract for $3.50 per bushel.[5]

Meanwhile, another contract, negotiated with the Alabama Manufacturing Company, which was also operating at Saltville, promised the state 50,000 bushels at the same rate, $1.75 a bushel, with an additional 50,000 bushels assured Alabama citizens, for, if the governor refused this second allotment, it must be sold to Alabama consumers at five dollars a bushel.[6]

In quite another direction, Governor Shorter was reaching for a supply of the all-important commodity. Colonel Bradford was sent to the Avery mine in Louisiana, apparently with a fund of $20,000 with which to purchase salt, of which sum he expended $6,447 for 155,950 pounds. He paid four cents a pound up to the amount of 100,000 pounds, four and a half cents for the remainder.[7] He secured a proposition from Mr. Avery for Alabama to supply the labor to mine her own salt at New Iberia at the price noted above, which proposition was promptly accepted by Governor Shorter.[8] This opening was followed up by his successor, Governor Watts, who also sent an agent to the mine the following spring.

The latter succeeded in bringing back to Alabama 3,000 bushels of salt, and in securing favorable terms for mining to an indefinite extent, but the federal successes in the Mississippi Valley deprived the state of any benefit from the contract.[9]

Meanwhile, Governor Shorter had been embarking on direct manufacture by the state at her own wells, since it was apparent that the yield from the operations of Figh and Company would be inadequate. In May, 1862, he appointed Mr. McGehee salt commissioner, who, after visiting the Virginia wells for observation and study, repaired to the state reservation where he proceeded to a thorough exploration of the region with a view to its utmost development. Innumerable obstacles had to be overcome at every step. Finally, however, he succeeded in establishing works and in getting them in operation by October, 1862.[10]

The action of the executive in entering the business of salt-making was indorsed by the legislature of this state when it met in the fall of 1862.[11] Although the governor recommended that if private parties proved unwilling to accept reasonable profits, the state should take possession of all the works and operate them for the common benefit, that step was never regarded as necessary in Alabama, as proved to be the case in Virginia.[12]

Mississippi entered the market for the purchase of salt in the second summer of the war. Governor Pettus sent agents to Virginia, Alabama, and Louisiana almost simultaneously. The missions proved, however, unsuccessful in the quest, so far as Virginia and Alabama were concerned.[13] The state of Louisiana was thoroughly scoured in September, 1862, by Pettus's agent, Augustin Chien, who felt that the salines in the northern part of that state could not do more than supply the people of Louisiana; and that the prospect was "gloomy indeed," if Mississippi were to rely on the Iberia mine, because the work was not being actively pushed. Furthermore, the prices demanded by the owner were ruinously high, three cents a pound or fifteen to twenty dollars per sack in Mississippi; and the line of transportation by devious river routes, including the crossing of the Mississippi, was seriously threatened by the foe.[14]

About a month later, despite the earlier discouraging report, we find another agent, Captain D. S. Pattison, dispatched to the mine with $20,000 and with the steamboat, *Newsboy,*[15] to purchase salt, or, in case he failed, to afford transportation for their salt to Mississippians who had succeeded in making purchases for family consumption. Fortunate in his quest, he started back with 40,000 pounds but was detained by confederate authorities on the Bayou Teche[16] for fear of federal gunboats until he was actually blockaded. With great difficulty he finally got the salt delivered in Vicksburg and ultimately distributed to destitute families. The agent was, of course, forced to abandon a contract with Judge Avery for 150,000 or 200,000 pounds of salt. It was the agent's conviction that if, contrary to his belief, the confederacy were able to hold the country, Mississippi should begin to mine her own salt at the cost of two and a half cents a pound, and that negroes, wagons, and equipment would have to be sent to the works from Mississippi, as no reliable labor was procurable in Louisiana.

Governor Pettus dispatched H. O. Dixon to Virginia to make contracts for salt, with the thought also of establishing furnaces for the manufacture of salt on state account. This ambassador found upon his arrival at the works in mid-September that salt was for sale only in very small quantities and at exorbitant prices, whereupon he promptly concluded that he must embrace the alternative of state manufacture. He found the proprietors reluctant to undertake negotiations, as the Virginia legislature was considering taking the salines into its own hands, but the proprietors promised Mississippi a supply next after Virginia. While awaiting legislative action, Dixon made two conditional contracts for 600-800 bushels per day at a cost of three dollars per bushel, including the tariff on the brine, which the proprietors had agreed to let him have at the rate of seventy-five cents per bushel of salt made. Finally, instead of seizing the works, as he was empowered to do, Governor Letcher of Virginia negotiated a contract for a supply of salt for his state which would consume all the brine of the existing wells. The proprietors, however, offered to allow Dixon to sink a well on the Preston estate, if Mississippi bore half the expense, with the right to make 200,000 bushels of salt a year, for which she would pay seventy-five cents per bushel, the Stuart Company securing the residue of the salt water. Because it was near the packing season, Dixon carried the proposition back for Governor Pettus to consider, but the latter did not feel justified in making the expenditure involved.[17] The governor then authorized Strong, Cunningham, and Company, a Mississippi firm, to manufacture salt at Saltville on private account for the people of northern Mississippi.[18]

The agent sent by Governor Pettus, probably in early September, 1862, to survey the possibility of a supply from Alabama was C. M. Vardin, who was to select a suitable location in that state for Mississippi's salt activities. In Washington County he found nothing worthy of such an undertaking, and so repaired to Clarke County where he reported an inexhaustible quantity of brine and lands in private possession favorably situated for such operations. Among such owners he discovered one truly patriotic man who offered Mississippi the use of his land gratis for this purpose.[19] Since the state would have to bore on this higher ground some seventy-five to one hundred feet, the agent found that the wells would cost on an average about one hundred dollars. However, as timber was available for fuel on Alabama land near by, he urged immediate procurement of the claim to the thirty acres in order to test the practicability of the enterprise.[20]

Discouraged by such meagre results, the Mississippi legislature then renounced hope of a supply from the contract method and by a law passed in January, 1863, provided a complete system for the manufacture of salt on state account by mining or otherwise, within or without the state, especially with a view to aiding the indigent families of soldiers. It appropriated $500,000 to carry the act into effect. Provision was made for a general salt agent who was to direct the manufacture and distribution, and who was empowered to appoint one or two skilled manufacturers, the agent being authorized to erect necessary buildings and to provide for the equitable distribution among the counties through duly appointed county agents.[21] This did not, of course, preclude purchase of the commodity where possible.

W. C. Turner,[22] who was sent, in accordance with the act, to the Alabama works in February, 1863, to buy, make contracts for salt, or, failing all else, to erect furnaces and manufacture on state account, made a contract with Norman and Company for 20,000 bushels at six dollars a bushel, 4,000 bushels deliverable July 1, 1863. Payment was to be made partly in provisions—corn, fodder, bacon, molasses, and sugar—and partly in money advanced to help erect the works.[23] After failing to meet their first installment, they asked on August 15 to be allowed to refund the money and to be released from the contract as it was out of their power to fulfill it. They proposed to sell Mississippi two bored wells and their two furnaces, which yielded daily ten to twelve bushels of salt.[24] Turner recommended the annulment of the contract and the taking over of the wells as a means of getting salt, and regarded their purchase as a bargain, since each would sell readily for from six to eight thousand dollars. These wells with one other within one hundred yards, which he had purchased, should, in his opinion, by the next October yield fifty to one hundred bushels of salt at less cost than to buy it.[25] Future purchases followed in 1864: in March, Mississippi bought the right to locate salt works and buildings and the use of the timber on 240 acres of land for $3,600 of a certain Henry Atchison;[26] in July it secured for $8,000 the right to bore two salt wells at the Morgan Salt Works with building privileges on adjoining land; while in December a further investment of $4,000 gave the state the right to bore one more salt well.[27] Hence the close of that year found the state of Mississippi possessed of several salt wells, furnaces, and 480 acres of timber land at the Alabama salt wells, with a large array of work-shops, cabins, offices, sheds, tanks, and other equipment.

Georgia was first in the field at the Virginia works and hence secured better terms for the brine there than any other state or company. No works were pressed forward with more energy. As soon as the Virginia legislature had decided in May, 1862, not to purchase or lease the works at Saltville, Governor Brown sent John W. Lewis, a Georgia confederate senator, who had proffered his services gratis, to Virginia as agent for Georgia. He succeeded in closing a contract with the proprietors for the use of water sufficient to make five hundred bushels of salt a day for the period of the war and for three months thereafter. He very soon turned over his contract and equipment to Major M. S. Temple of East Tennessee to manufacture the salt. The cost of the article to the state was estimated at $1.50 per bushel, as weighed from the kettles.[28] It was the governor's intention to sell the remainder, after the wants of soldiers' families had been met, to the inhabitants of Georgia at cost of production and handling.[29] By the end of June, Temple had begun erecting furnaces and by the close of October was meeting his stipulated quota of five hundred bushels a day. By the following March he was representing that the price agreed on was utterly inadequate, owing to the increased price of labor and provisions. Governor Brown, fearing bankruptcy on the part of the contractor and the consequent suspension of the work, recommended to the legislature at its special session in April, 1863, an alteration in the contract on condition that the full complement of salt be delivered.[30] The legislature obligingly approved the increase and made it retroactive to January 1 by requesting the governor to modify the contract.[31] But even this liberality did not satisfy the Temple company, for by the following November they notified the state of their desire to discontinue making salt, whereupon the Seago, Palmer Company of Atlanta stepped into the breach with a contract with Georgia for the same amount of salt, each party to pay one-half the cost of transportation and each to furnish one-half the sacks.[32]

By the following December the output of salt had been greatly increased, despite embarrassments, so that the official works had shipped to the state 150,000 bushels. The supply for the state even reached the estimate of a senate committee of 275,000 bushels with the supply yielded by other works in Virginia. In fact, the Salt and Iron Supply and Transportation Committee felt moved to congratulate the legislature that though there was scarcity in salt in some portions of Georgia, the general supply was in good condition.[33] But this rosy view is not borne out by the evidence.

Meanwhile, Governor Brown had neglected no opportunity to secure the precious commodity. He agreed to take all the rock salt from New Iberia that a certain firm could deliver from October 28, 1862, to March 1, 1863, at $7.50 a bushel.[34] In Georgia, despite the confidence manifested in the fall of 1863, the price went up; in the spring of 1864 we find Governor Brown contracting with Bigham and Cox for one thousand bushels a month to be delivered in Atlanta or Augusta, not at $7.50, but at ten dollars a bushel, the state bearing the cost of sacks or barrels and agreeing to meet any excess of transportation costs not covered by the price stipulated.[35]

Out of a larger sum set aside for indigent families, Arkansas designated $200,000 as an amount with which salt should be purchased for the families of soldiers within the counties overrun or occupied by the public enemy.[36]

Governor Milton of Florida, discovering the convenience and advantage of purchasing salt in quantity for the benefit of soldiers' families, devoted a portion of the amount appropriated for their relief to the purchase of salt, which was distributed in lieu of that which would have been purchased by them at a much greater price. He felt so gratified by the results that he recommended a specific appropriation for the purchase of salt for distribution in those counties where it was most difficult to be obtained and hence most costly.[37]

Nowhere is the story of salt-making more interesting than in the record of the activities of North Carolina, which were conducted in two places: on her own coast, and in the neighboring state of Virginia. She wasted no energies on an effort to buy in the open market,[38] but set about at once creating her own supply. Since she was without adequate salt springs within her own borders, the subject obtruded itself upon public attention almost at once. It entered into the convention of December, 1861. That body straightway passed an ordinance providing for the election of a commissioner to manufacture salt and to sell it to the citizens of the counties at cost. The commissioner was vested with the powers necessary for the discharge of his task by this ordinance and by an amendment passed before the convention adjourned: the right to employ necessary agents and laborers; to procure necessary materials, for which funds in the amount of $100,000 were provided; the right to buy or lease land for erection of salt works, and, if necessary, to seize it under the right of eminent domain; with the right to bore for salt and establish works wherever he deemed expedient; and the right to employ or impress free negroes on the works with the rations and pay of soldiers. All persons employed in salt-making under contract with the salt commissioner were declared exempt from state military duty.[39]

The salt commissioner was designated at once, Dr. John Milton Worth, later succeeded by Daniel G. Worth, his nephew and a son of the state treasurer.[40] Salt works were first established at Currituck Sound, the best locality on the North Carolina coast, but the expenditure was very promptly to be written on the debit side by the loss of Roanoke Island. Morehead City was then selected as the next best locality. The state works were just beginning to produce salt when the fall of Newbern through inadequate numbers of troops, resulted in the capture of the pans and works by the enemy, so that the commissioner was thrown back to a ground start again. The third point selected was a spot eight miles from Wilmington, where the enemy's war vessels could not approach, some twenty miles from the forts at the mouth of the river. The sea breezes made officials regard the location as pleasant and healthful; indeed, it was the location of many summer residences.[41] Private manufacture of salt was practised here more extensively than at any other place in the state. Here the salt commissioner ground out salt from the sea for two years at a number of different places.

The two war governors, each in turn, urged on the work to the limit of his ability. Governor Clark readily acceded to the commissioner's request for power to impress free negroes,[42] and wrote urgently in August, 1862, "Put everything to work and make salt and you will receive the thanks and commendations of the people. But if you fail and private parties succeed in making it you will be censured and probably suspected. So let me urge you to spare no labor, money or pain in making salt." [43] In similar strain Governor Vance wrote frankly in the following October, "It (Worth's report) is satisfactory in all respects, except the amount produced. In the present Emergency it is desirable to have salt without regard to Expense. The vast amount of meat that will be lost without the salt renders the price of it a small consideration." [44]

Worth had just reached the point of producing 250 bushels daily when the work was interrupted by a virulent type of yellow fever which raged two months from the middle of September to the middle of November, 1862, causing 441 deaths out of about 1,500 cases, and almost entirely suspending the salt works. The epidemic had been introduced by a steamer from Nassau in July, but public attention was not directed to the death of two of her crew until the disease had spread and thrown the city into a panic.[45] Although there was no recurrence of this epidemic, the commissioner had to contend at various times with chills and fever among his hands so that in September, 1863, over half of his men were ill and recovering slowly; again in November, 1863, he lamented illness which hampered the work; and in October, 1864, he again complained of having sixty hands unfit for service.[46]

Worth found himself obliged steadily to increase the price of salt to the counties on account of the enormous advance in cost of provisions and materials. In September, 1863, he charged six dollars a bushel,[47] seven dollars in November; nine dollars in December, ten dollars in February, 1864; thirteen dollars in May and fourteen dollars in September, a sum which he declared even then did not cover the cost.[48]

The conflict between the civil power of the states and the military was nowhere more sharply illustrated than in the long-drawn-out duel at Wilmington in regard to salt-making by the state of North Carolina. The fate of the state salt works was in doubt from January, 1863, to November, 1864, during which period one crisis followed another. Early in January, 1863, the military authorities impressed the steam-boat and flat-boats, by means of which Worth ran wood and brine through the sound to the river-side works, thus reducing the number of teams.[49] About the middle of the month all his teams and hands were impressed and put to work, making bridges and breast-works for the military, while at the same time he was expected to supply rations to the teams and men, thus exhausting his supplies without leaving him power to produce the salt for which they were exchanged.[50] He, therefore, reluctantly allowed some of the teams to go home, though aware that a military attack might prevent their return.[51]

In less than two months the commanding general, casting jealous eyes on the salt commissioner's sound men, called on him for 120 men, claiming that the commissioner had more than he needed.[52] Though an inspection by a citizen of Wilmington and by an army surgeon sustained Worth, the strife went on,[53] for to jealousy was added a conviction in the mind of the commanding officer, General Whiting, that many of the salt-makers were traitorously supplying information to the enemy.[54] The situation reached a climax in the judgment of Whiting in April, 1864, when the enemy landed at one of the state salt works, setting it afire and carrying off a number of conscripts. The federals did not do a great deal of damage, as the shells thrown into the furnaces did not break the pans, while the sheds over the works, dampened by the steam from the boiling kettles, burned so slowly that the fire was soon extinguished, damage being limited to the pump and engine. Worth declared that if he could pump water, "two-thirds of the works could be run," [55] though the blacksmith shop and stables had been entirely consumed. The entire damage was estimated at $15,000. Whiting ordered the salt works removed to the Cape Fear River, although Worth thought they might better be entirely discontinued than removed to that point. Though Governor Vance denied the general's authority to control a state work, and directed Worth to repair and continue the works, Whiting issued a second peremptory order, when he learned that the works were still running, for the suspension of all salt works, state and private, on Masonborough Sound. He furthermore impressed the three "flats" by which Worth set such store.[56] That Whiting had his eyes fixed anxiously on the conscripts is evident from the observation in his report to the adjutant-general that the conscripts "should be turned over to the camp of instruction." [57]

The solicitude of the father for his son comes out strongly in certain communications of Jonathan Worth. He wrote a long exposition of the whole controversy between General Whiting and the salt commissioner to W. A. Graham, a member of the confederate senate, on June

8 and one to the governor at about the same time. He advised his son not to obey the order, even if he were put under arrest, until the governor was heard from. His real motive, to save his son from the ranks, is clear in the following sentence: "If the Govr should yield, still he has no power to remove you and there would be much for you to do for some months to come in winding up the business." He blamed the governor for preoccupation with the approaching election and for making little effort to prevent "the calamity" to the works.[58]

Once more Vance came to the rescue of his salt commissioner, for by direction of General Beauregard, General Whiting was ordered to suspend his action until Governor Vance could be heard from, while the enrolling officer was directed a few days later to suspend enrollment of the employees. When Vance was heard from late in June, it was through a most vigorously worded protest against Whiting's action, addressed to the secretary of war. "Before submitting to this outrage on the rights and dignity of the state, as well as its vital interests, I deem it proper to appeal to you, in the hope that this officer can be made to do both what is right and proper. If he can, then I have no objection to his remaining in command at Wilmington. Otherwise, I shall be compelled to ask his removal." He pointed out that a change of location would cost $60,000-$70,000. Governor Vance was betrayed into a rash offer to provide a guard of state forces to protect the place against disloyal workers, if the central government were unable to do so.[59]

Whiting grudgingly conceded that the work might go on until the secretary of war gave a decision, though it was only a few weeks later that he declared to Governor Vance that he had proof of many salt workers belonging to a treasonable organization called H.O.A. (Heroes of America). Meanwhile the flat-boats were being rendered useless by the mining of the channels of the sound with torpedoes and obstructions.[60] Governor Vance found himself somewhat embarrassed when General Whiting on July 20, 1864, through the intervention of Secretary Seddon, accepted the offer of a state guard to protect the works.[61] After a delay of more than a month Vance found it so difficult to furnish the two companies required by General Whiting that he raised the question with the salt commissioner whether the works could be removed to Confederate Point or whether the disloyal workers could be singled out and removed, to neither of which propositions could Worth send an encouraging answer.[62] Meanwhile, the flats were suffering deterioration, as exposure for months after their seizure had led to their being devoured by salt-water worms, so that they would require reconditioning before again being put into use.[63]

Finally, at the long last, Whiting had his way with his superior officers[64] and the blow fell. On October 27, all private salt works on the sound were notified to suspend operations. On November 15, quite suddenly, Worth was ordered to remove from the works before sundown all teams, movable property, and men, as after that hour only troops would be allowed outside the city lines. Salt employees were to be sent to enrolling officers.[65]

Governor Vance, seeking another location for the state works, directed Worth to search the coast of South Carolina for a suitable location, but the latter held all conditions there unfavorable, especially the lack of bricks for the furnaces and the difficulty of shipping them from North Carolina.[66]

In reporting the action of the military to the assembly, Governor Vance condemned the action as unwarranted. Apprehensive of the menace to the salt equipment from a possible federal attack, he ordered the property removed to the railroad. He seemed unable to offer constructive suggestions for future works, and contented himself with caustic remarks about his intention to re-erect the works if a spot could be found "where salt was not held in such disesteem and its manufacture would not be deemed injurious to the cause of the confederacy." He asked the advice of the assembly.[67] That body seemed no more resourceful than the governor and relieved its feelings by a futile declaration that the employees of the state salt works were as entitled to exemption as any state officer, and instructed the governor to use the militia to prevent interference with them or with the salt works. It hurled defiance at the central government by declaring guilty of a high misdemeanor anyone who should seize the flats, mules, wagons, or property of the state, or interfere with the salt works.[68] Later in the session a resolution authorized the transfer of the superintendent and of the property used for salt-making at Wilmington to the salt works at Saltville or to any other point held advisable by the governor.[69]

Meanwhile, state activity by North Carolina had been going on simultaneously but also with difficulties in Virginia. Upon the strength of a rumor that Stuart Buchanan and Company were offering the governors of neighboring states the opportunity to make salt,[70] Governor Clark on June 24, 1862, appointed N. W. Woodfin and George W. Mordecai commissioners to go to Saltville to make contracts for the purchase of salt or for the use of the salt-works.[71] A contract was promptly negotiated on June 30 for salt water in sufficient quantity to make 300,000 bushels a year for the period of the war, together with the right to erect furnaces and necessary buildings. The charge was a toll of seventy-five cents per bushel for all salt made under the contract, an increase of twenty-five cents over the charge to Georgia. North Carolina was bound to begin erecting furnaces within thirty days. The Stuart Company was obligated to spend up to $25,000 in boring for brine and to pay the cost of the furnaces and improvements if they failed to supply the water. Priority of claim upon the brine was definitely fixed in the following order: Stuart, Buchanan Company, the state of Georgia, the Planters Salt Company of Georgia, and North Carolina.[72]

Woodfin consented to act as superintendent and state agent at the works, in which post he continued until the close of the war. For the next six weeks after the contract was made he was busy, energetically pushing his undertaking, buying lumber for a tank, bricks for a kiln, negotiating for kettles and salt pans, soliciting aid of the confederate ordnance department in procuring a supply of iron for kettles, and seeking laborers.[73] Governor Clark contributed to the work by advancing $35,000 from the treasury, by sending between twenty and thirty tents, by publishing a notice to citizens of the need of supplying labor, and by urging Woodfin not to hesitate at price of materials or labor.[74] By November 17, when Vance reported to the assembly, he stated that there were 200 kettles in operation, producing over 1,200 bushels of salt a day.[75]

Inasmuch as the original legislation on the salt question, that of the convention of 1861, provided only for works within the state, fresh enactments to legalize the activity in Virginia were necessary. The assembly promptly confirmed the contract with Stuart, Buchanan and Company; the sum given Woodfin was appropriated; the office of superintendent at Saltville was authorized; and, finally, recognizing the stringency which existed at the time in regard to salt, the governor was authorized to purchase 100,000 bushels of salt to relieve the scarcity and to purchase and conduct any salt works then in operation.[76]

Despite constant hindrances and almost insuperable obstacles, Woodfin pushed forward the production of salt. By the end of February he was able to state that in five days he had loaded five cars and expected very soon to run 236 kettles, turning out 1,500 to 1,800 bushels a day. By July 1, 1863, he had shipped 86,729 bushels and had 20,000 bushels on hand.[77]

By September, 1863, Stuart, Buchanan and Company desired to change the contract so as to be paid for the brine furnished, in part at least, in salt instead of in depreciated confederate currency. When Woodfin called as usual on the first day of the month to pay the rent for the brine, the proprietors refused to accept confederate money for the salt shipped in August.[78] He expressed the hope that the confederate government would take possession of the property at Saltville and was irritated into declaring of the proprietors, "I think if they owned all the world they would want ground enough outside to put up a little furnace to make a little more salt in so as to try and make a little more money." [79] Since Woodfin was without power to alter the contract, the controversy dragged two months until he agreed, subject to ratification by the governor or legislature of North Carolina, that a board of three disinterested arbitrators be organized to whom should be submitted the question of what allowance should be made for the depreciation of the currency, the award to cover only the next three months.[80]

In April, 1864, a new contract was negotiated between Stuart, Buchanan, and Company and North Carolina for the direct purchase of 54,000 bushels of salt, 6,000 bushels deliverable per month from April to the end of the year in car-load lots of 3,000 bushels at a time. Payment was stipulated at six dollars per bushel in the new confederate issue of April 1, 1864; until the new issue was available, the state would pay in certificates of deposit or in four per cent bonds at 75 cents to the dollar, if preferred by the proprietors. The company agreed to keep a furnace running constantly to supply the state with salt while the state in turn agreed to deliver to the company 500 bushels of corn each month as nearly as practicable, the cost of the corn to be deducted from the price of the salt, and the company to deny delivery of the salt in exact proportion to the failure of North Carolina to provide the corn.[81]

The suspension of the works at Wilmington, late in 1864, by General Whiting made Governor Vance and the legislators look the more anxiously to Saltville, where the situation was not improved by the absence of Superintendent Woodfin for a number of months, from March until late September, because of personal considerations.[82]

The legislature at its session of 1864-65 duly confirmed the contract with the Stuart Company for the purchase of 50,000 bushels of salt, setting aside $150,000 for the purpose, and appropriated $250,000 to purchase an engine and train to transport salt from Saltville to North Carolina.[83]

On the very eve of the surrender of the confederacy, March 8, 1865, we find Woodfin negotiating yet another contract with the Saltville proprietors in order to assure the state salt for the year 1865. He arranged for the outright purchase of 48,000 bushels of salt at eight dollars per bushel, though provision was made to cover the damages of fluctuation in the confederate currency. It was to be delivered in car-load lots of 6,000 bushels for eight months from May 1, North Carolina to furnish the sacks or barrels and the twine to sew the sacks. Woodfin agreed to pay $200,000 in advance, but, since the contract was to terminate with the war, he stipulated that repayment for undelivered portions should be made at the rate set in the contract.[84]

North Carolina continued its manufacture till the fall of the confederacy as best it could under the increasing difficulties. As late as March 2, 1865, it bought out the interest of the Alabama Salt Manufacturing Company in 102 kettles and in two furnaces and agreed to pay the Stuart, Buchanan Company $20,000 in confederate states treasury notes before April 1 for the use of the furnaces when repaired and for delivery of brine until January 1, 1866.[85]

From South Carolina came evidence of vigorous, if ultimately rather futile, efforts during the period of acute scarcity in the summer of 1862 to meet the salt famine in that state by purchase. By action of the Executive Council in February and May, 1862, contracts were negotiated with nine different producers, whose works were all located, with one exception, within the state. The contracts called for varying small amounts of salt, ranging from 500 to 800 bushels, deliverable before the middle of August. The price stipulated was in most instances four dollars a bushel, though a few contracts called for only three dollars, and one was fixed at $4.25. A few arrangements were indeterminate, demanding all the maker could produce or over half the amount made. The sad part of these contracts is the fact that upon close scrutiny it appears that only 666¾ bushels in all were actually delivered to the state by August 13, and that only 1,598 were even manufactured—and that by one firm. The Executive Council tried to stimulate the manufacture of the article by setting aside $50,000 for assistance in its production.[86]

Governor Pickens that same year endeavored to provide a supply of the necessity for South Carolina by seeking a contract with the owners of the Virginia works, but he found the problem complicated by the fact that conditions which accompanied the proposals bore upon the private rights of a distinguished South Carolinian holding a mortgage on the works. This fact precluded him, on grounds of justice, from accepting them. He next appointed a man to make arrangements for the transportation of 100,000 bushels, but found this impracticable, even if such an amount of salt were procurable in time for the packing season, which he regarded as highly unlikely. Hence, for the season of 1862-63, this state was forced to depend almost entirely upon the production of her own coast.[87]

Following the governor's statement, the legislative assembly worked out a plan which had unique features. It empowered the governor to contract for 20,000 bushels of salt a year from private manufacturers for a period of twenty years at fifty cents a bushel, under conditions which required production of the salt by solar evaporation within the state; purchase from any one person or establishment was limited to 5,000 bushels annually; and delivery was permissible at Charleston or Columbia at the option of the seller. The governor could dispense with the delivery of contract salt as long as the current price in Charleston rose above fifty cents a bushel. Provision was made for the purchase or erection of two public magazines for the storing of the salt, each to be capable of holding 100,000 bushels, one in Charleston, the other in Columbia. Three commissioners for each magazine were to receive the salt as it was delivered under the contracts, and have it stored in the magazines. Whenever the current price at Charleston or Columbia rose above $2.50 a bushel, the commissioners were to sell to citizens of the state at $2.50 a bushel under rules approved by the governor, which were designed to secure to all parts of the state a fair share in the benefit of such salt and to prevent opportunity for speculation or extortion. The magazines were not to be commenced until one year after the passage of the act.[88] The law was cleverly drawn to "peg" the price of salt to the consumer at about $2.50, and to insure that it would cost the state less than that sum.

Other states which engaged in the manufacture of salt, though only to a slight degree, were Louisiana[89] and Texas. The legislature of Texas authorized the governor the first year of the war, November, 1861, to appoint an agent to take possession of the famous salt lake, *El Sol del Rey,* to sell salt at the customary price, unless deviations were authorized by the governor, and to collect and pay the revenue to the treasury.[90] Two years later the Military Board was authorized to manufacture salt at the salt springs and creeks near the Double Mountains and at any other place where it could be made advantageously, the sum of $50,000 being made available for the purpose.[91] It does not appear that the Board availed itself of its authority. Virginia entered the field of salt production on a grand scale, but so important and far-reaching was her action as salt manufacturer that it must be reserved for special treatment in another chapter.[92]

It was not enough that the several states engaged in the industry of salt-making; certain counties embarked upon it as a county venture. It is not possible to say just how many, but there is enough evidence to show that such ventures existed in several states. One county only in Alabama, so far as the writer has been able to ascertain, embarked early on the policy of salt production for its families, Clarke County, in which the chief wells were located, though no record has been found of the amount produced. In September, 1862, the county appropriated $2,500 to construct a salt factory and $600 to purchase requisite equipment.[93] The county records indicate activity up to 1864, for sums were ordered paid that year for mules bought for the county salt works and for fodder furnished the works.[94] An act of the Alabama legislature encouraged the courts of other county commissioners to engage in the manufacture of salt for their indigent families. Authority was granted them to buy or lease furnaces and as much land as necessary, to make contracts, and to provide salt at the cost of manufacture and expense of delivery; furnaces erected on the state reservations were to pay no toll to the state. Any surplus salt above the needs of indigent families might be sold at market price to residents of the county. A special tax, up to one hundred per cent on the state tax, was granted the county commissioners, if necessary in order to raise funds. Naturally, if the county provided an adequate amount of salt for its poor families, they were not entitled to any from the state appropriations.[95]

It would appear that a large number of counties made contracts for the purchase of salt directly from the Stuart, Buchanan Company at Saltville. Shenandoah County of Virginia and Banner County of North Carolina[96] are cases in point, though no less than sixty Virginia counties made such contracts during the first half of 1862.[97] And at least one Virginia city, Petersburg, bought through its city council.[98]

CHAPTER VII

# SALT OFFICIALS OF THE CONFEDERACY AND THEIR HARASSING PROBLEMS

JUST AS in the World War strange officials appeared, such as Food Commissioner, Fuel Commissioner, and Railroad Commissioner, so in the confederacy officials unknown before, such as salt agents or salt commissioners, are encountered. The method of distributing the commodity so as to make it available for all, but especially so as to help the poverty-stricken soldiers' families, was similar throughout the various states of the confederacy. In general, some one was put in charge for the entire state, the commissary-general in Georgia, the quarter-master general in Alabama, a general salt agent in Mississippi [1] and Arkansas. In North Carolina the two producers of salt for the state, the salt commissioner at Wilmington and the salt agent at the state salt works at Saltville, Virginia, both shipped directly to the county salt agents. This state-wide official shipped the salt to local distributing agents, sometimes directly to the county salt agents, sometimes to a limited number of distributing points, from which the county agents of the particular section drew their allotments. Georgia and Alabama were examples of states which used the intermediary system. In the former, Commissary-General Whitaker shipped to nine centrally placed depots in the state;[2] in Alabama Quarter-Master General Green sent supplies to six distributing centers.

Whatever the method by which it reached the counties,

it was there parceled out to the families according to need, with preference given to indigent soldiers' families, by county salt commissioners, by justices of the peace, by county agents, or by justices of the inferior court.

In Georgia, in the first distribution, one-half bushel was to be given without charge to the widow of each soldier who had died in service, the justice of her claim being based on lists compiled by the justices of the inferior courts; one-half bushel was to be sold to the family of each soldier in service and to each widowed mother of a soldier, at one dollar. The claim was established by certified lists. The justices of the inferior courts of the counties were expected to advance the money from the county treasury to pay for the salt before it was shipped, and to bear the expense of its distribution, which sums were refundable to the treasury from the returns of the sale of the salt.[3] A soldier's widow or wife who wanted more than one-half bushel for her family use might purchase in preference to all others at the price paid by others. All other heads of families must pay at the rate of $4.50 a bushel, purchases being restricted to one bushel till all had received that amount. Purchasers were required to bring their own sacks, and all sales were for cash, as agents were required to remit weekly. Distributions were made to the counties in the order in which the courts presented their reports.[4] The governor furthermore directed that in the process of the distribution the northeastern part of the state receive first consideration after the soldier's families, as that section was too poor to have many slaves, had furnished large numbers of soldiers to the service, and was remote from the railroads.[5] There were at least three such distributions in Georgia to soldiers' families during the war.[6]

In Alabama, a sharp distinction was made in the disposition of the salt derived from the lessees, J. P. Figh and others, and that made at the state salt works. Indeed, the shipments from the respective works were branded as follows: the words *County Commissioners* were stamped on all salt from the lessees' works, and *Alabama* on the salt from the state works, while separate accounts were carefully kept. The salt for the counties, under the law which required the lessees to turn over to the state two-fifths of all the salt they made as rental for the land, cost the state nothing for its manufacture, and hence was distributed in the counties through the courts of county commissioners to the families of soldiers at such "just" price as the governor might fix, merely to cover the expenses of handling, sacking, transportation, and the salary of a resident agent at the works. The first regular relief shipments of lessee salt were made at the rate of twenty bushels to the county during September, 1862, and were designed for the indigent families of volunteers.[7] The first regular state-wide distribution did not occur until January, 1863, and then the uniform price of $2.50 per bushel of fifty pounds, as required by the law, was charged. Commissioners were expected to pay for the salt for indigent soldiers' families from the sum received from the state for their support and to distribute it gratuitously.

The state-manufactured product and that purchased was, on the other hand, distributed and sold to the inhabitants at prices not exceeding the cost of manufacture or purchase and handling, the object being, not speculation, but the furnishing of salt at cost to the consumer, with due regard to the wants of counties remote from means of transportation. In this case the salt commissioner, naturally, was the person designated to fix the price.[8] No one was allowed to buy except for private use nor in quantities greater than twenty-five pounds per capita until all the people had been supplied with this initial small allotment.[9]

The price varied on the product derived from the lessees within short time limits. The governor directed Figh and Company to send the first installment to Washington County on September 1, 1862, for $1.25 a bushel; September 22 he fixed the price at $2.25 and carrying charges to Barbour County; and in January, as we have noted, the price was set at $2.50 a bushel. The first allotments sent out to the counties in July, 1862, as an emergency measure, were secured from a stock which the state purchased at high prices, and which, therefore, was parceled out in pecks and half-bushels at ten cents a pound net at Selma.[10] Governor Shorter worked out a complex plan whereby he hoped to supply the fourteen northern counties from Saltville, Virginia, at the same price as that charged the other counties for Alabama salt.[11] Later in the war, it was decreed that salt acquired in any manner should be distributed among the several counties to indigent soldiers' families at a uniform price to be fixed by the quarter-master general each fiscal quarter so as fully to reimburse the state. Legislation explicitly excluded families of deserters from the benefits of the act, but no discrimination was to be practised against poor families of soldiers based on degree of destitution.[12]

Mississippi, Arkansas, and North Carolina used a more direct method of distribution; the two former dealt directly through a general state agent with county agents; the latter allowed her two manufacturers of state salt, the salt commissioner at Wilmington and the salt agent at Saltville, to ship to the county agents. In Mississippi and North Carolina the county agents were selected by the justices of the peace or boards of police of the county, but in Arkansas the state agent selected the local agents.

Mississippi did not provide a general salt agent until April, 1863, when Governor Pettus appointed the quarter-master general, Colonel West, agent to receive and distribute to the boards of police of the various counties the pro rata share of all salt obtained by purchase or manufacture. The Mississippi and Arkansas state agent had power to establish a general depot to which all salt was to be shipped for distribution after manufacture. This official then sent to the county salt agents appointed by him or by the boards of police of the counties a pro rata share, according to the number of soldiers and their dependents, based on lists furnished by the probate clerks of the counties of Mississippi and by the county agents in Arkansas, at a charge covering cost and transportation.[13] The boards of police were required to furnish salt, first to soldiers' families at cost; then to furnish any family one year's supply for its own use; and finally to sell the remainder to the most needy persons of the county at cost.[14] Naturally, deviation from the strict pro rata rule occurred, as when, to prevent the salt from falling into the hands of the enemy, the general agent issued it to such counties as could send for it or to citizens for their own use; or, again, as when a county having lost its quota for indigent soldiers' families by capture or destruction of the salt by the foe, a second installment was issued from the salt depot free of charge.[15] By April, 1865, Mississippi had been able to make but two general distributions, the second sometime in the fall of 1864.[16]

North Carolina was conspicuous in that there does not seem to have been set distributions nor quotas, despite initial instructions from the governor to divide the salt among the counties according to population. Shipments being made as the two agents had salt available, there were great discrepancies in amounts among the various counties. In order to equalize the distribution the county commissioners exchanged with each other the salt due from Wilmington as against that due from Saltville in order to capitalize for a county on the coast its relative nearness to the state works at Wilmington.[17] Furthermore the individual counties seem to have been expected to meet the expense of their shipments—and more, to pay it in advance,[18] though such a rule, naturally, could not be rigidly adhered to.

The system of distribution which was worked out in Virginia was very complicated.[19] A state agent at Saltville was to receive the salt from the manufacturers and to ship it to the depots established in each congressional district where a subordinate state agent would receive it. Each county or corporation must advance the amount for the salt, which was to be allotted according to population, to a duly appointed county or corporation agent, vested with authority to buy the salt at the depot from the state agent. The county or municipal agent became responsible for the distribution to the individuals in the counties or cities. Priority in receiving allotments was established among the districts by lot and the amounts fixed according to population, including refugees.[20] The law of March 30, 1863, legalized the governor's action in regard to distribution and placed the matter under the direction of the Board of Supervisors. Upon recommendation of any three or more of the justices of the county, or of the state senators and delegates, agents were to be appointed by the county and corporation courts, or by the Board of Supervisors, if the presence of the enemy prevented the court convening.[21] When distributions occurred, great numbers of the citizens came riding in from all the adjoining counties, in every conceivable sort of conveyance, a large majority of the vehicles freighted with produce which the owners hoped to sell to the soldiers.[22] The last installment of state salt in Virginia in December, 1864, which was sold at one-fourth the market price, allowed only six pounds per capita.[23]

From first to last there was a bewildering array of state officials concerned with the handling of salt. Beyond the general state official and the county agents concerned with the distribution, as just described, there was endless variety. For a state which embarked on salt production on a large scale North Carolina had the simplest system: two producers of salt who dealt directly with the county agents. Alabama had perhaps the largest array, ranging from Quarter-Master General Green to Commissioner McGehee, who presided over the state works in Alabama as salt manufacturer; from Captain Speed, who discharged similar duties at Saltville, to Long, who was superintendent of the Alabama state works; and from Dodsen, the state agent to receive the salt from the lessees, to Snodgrass, the special agent for transportation for the Virginia salt.[24] The relatively large group sent to various points in the frantic effort to locate a supply during the summer of 1862, when the stringency was first generally felt, should not be forgotten. In addition there were assistant salt agents and overseers at all the works.[25]

The duties of the man who was responsible for the manufacture of the salt were sufficiently arduous to insure the ample earning of his modest salary, not to mention the obstacles which naturally harassed him during war-time. The instructions laid upon McGehee by the vigilant Shorter [26] may be taken as typical. He was directed "with all possible energy" to develop a full supply of brine, to erect furnaces capable of producing 500 bushels of salt immediately with hint of the necessity of future enlargement; to secure an adequate supply of laborers, assistants, and agents; and to provide supplies for workmen and teams. He must direct and control all shipments of salt to fulfill contracts made by him personally, presumably for supplies; he must ship the rest of the product to the quarter-master general or up the Tombigbee River, as that official might direct, forwarding allotments by every boat, if practicable. He was further to lease the use of surplus water at both reservations, for one-tenth of the salt manufactured, exacting deliveries at least twice a week; as the governor's proxy, he was required to locate all persons engaged in the manufacture of salt for public or private use on the public lands, setting apart for each such maker suitable grounds in such manner as to protect the state works and the works of the chief lessee, Figh and Company, from interference; and he was to prevent collisions between persons making their own salt. He was directed to bring to the attention of the public the fact that without a lease no person was allowed to manufacture salt on the public lands with intent to sell. He was vested with authority to use the military power, and was required by executive order to report to the governor weekly; under the law he was held to quarterly reports to that official, and annual reports to the assembly. He was forbidden to sell any salt at the works, except in pursuance of contracts already made or contracts necessary to secure labor or supplies.[27]

Problems and obstacles which would naturally beset the path of any official undertaking to conduct a new state business enterprise were multiplied a hundred-fold by the war activities and conditions. For no state are the records more complete in this respect than for North Carolina; from them one can get a picture which clearly reveals that the path of the salt commissioner at either Wilmington or Saltville was not one strewn with roses. Glimpses here and there in other states show that life was no more pleasant or smooth for any other courageous soul who undertook to produce the commodity for his state.

First of all, he faced scarcity of equipment and of all the materials from which he was to construct his equipment. Although there were foundries to cast kettles in most of the cities and also iron works in northern Georgia, that state experienced great difficulty in securing the necessary metal, especially as it was felt that she should solve her problem without calling upon the confederate government for a supply of iron.[28] Woodfin experienced at Saltville some difficulty in securing compliance with the contracts he had negotiated for 30,000 pounds of kettles for the North Carolina works, some of which had arrived broken and hence had to be replaced.[29] He evidently built his furnaces of rock which he quarried himself, though his colleague at Wilmington erected his of brick which he had to make himself. Woodfin had to supervise masonry works, build shanties for the hands and storehouses for the salt and supplies, thus filling the rôle of carpenter, smith, and mason; he had to make and repair roads. Sacks and twine or barrels for shipping the salt could present themselves as problems. North Carolina required for her works in Virginia about one thousand sacks a month during the rushed season of 1863 for the counties unable to provide sacks, and applied to the quarter-master's department for them. A few months later the agent was trying to get "good Osnaburg" for sacks in order to secure this article for the people at the same price the state was paying instead of at speculator's prices.[30] About the same time, April, 1864, Commissioner Worth was finding it impossible to get sacks and was forced to use barrels exclusively.[31]

The one most essential element for successful production of salt was a steady, copious supply of brine, but this almost invariably failed the manufacturers at Saltville at times, as the original lessees were often unable to deliver the vast quantities of salt water which they had contracted to furnish. Hence, the agents frequently found themselves obliged to "blow out" their furnaces. Another interesting phase of the problem was the fact that if the brine were sent through the logs without first being exposed to the air to afford opportunity for the escape of gas, the latter accumulated in the bored logs and thus obstructed the flow of the water. Therefore, we find Superintendent Clarkson of the Virginia state works eager for a cistern.[32]

The problem of provisions for man and beast was one from which a salt-maker on a large scale was seldom free.[33] Woodfin faced first of all the dilemma of purchase only at high rates in exchange for salt. He found it necessary in the beginning to bring everything from North Carolina. He brought from the mountains, 193 beef cattle, which fed his hands for two and a half months, and left him some eighty beeves to put up in pickle for winter and spring use, which must subsist his force till spring brought new bacon. His problem was simplified by the fact that he procured the great bulk of his grain—corn, wheat, and rye—by contract with county commissioners in exchange for salt, counting the salt offered in payment as a part of the county quota, but giving such counties preference in order of distribution.[34]

His colleague in the job, Worth, had been drawing his supplies of grain and forage almost entirely from South

Carolina, but by the close of 1863 he was finding it almost useless to buy grain there because of the impossibility of securing delivery, since the railroads were taxed with transporting corn for general Lee's army. Corn purchased in April was still undelivered to Worth in December. Hence he had to decline corn at good terms of exchange for salt. That source of supplies was abruptly closed to him the following February by an order of General Beauregard forbidding further exportation of subsistence or forage from South Carolina except for military purposes. Thereupon Worth had to discharge one-third of his teams. He had to pay four dollars per bushel for corn and then haul it himself.[35] As the difficulties of transportation grew exceedingly serious toward the close of the war, the resourceful Woodfin conceived the idea of making a considerable portion of his provisions in the vicinity of the works, and employed an active business man in Smyth County within ten miles of Saltville to cultivate a farm and sell all the surplus at government prices.[36] Though Clarkson, director of the works for Virginia, advertised that counties furnishing certain supplies and labor should have preference in obtaining salt, even this inducement proved unavailing in bringing in food supplies.

Weather bestowed no special smiles on a man just because he was trying to supply his state with salt. At Saltville particularly, but also at the Clarke County works in Alabama and at those in north Louisiana, the winter season or the floods of spring brought almost, if not complete, cessation of the work. Roads became impassable in the mountains of Virginia so that salt-making was almost suspended from Christmas to the first of May, and, even if the works could be operated, no salt could be shipped out. The various state agents of the works there must perforce lay in a supply of perhaps 1,200 cords of wood per winter month before winter set in. In the other saline areas the early spring floods automatically halted the works for a number of months.

The supply of wood for fuel was a basic necessity which engaged the anxious thought of the commissioner or agent constantly and increasingly, as the timber in the immediate vicinity of the works became exhausted, when the question immediately merged into the wider problem of transportation. J. M. Worth rejoiced in May, 1863, that he had 2,000 cords of wood on hand for the next month, but did not fail to note that by fall he would have cut down all wood within a radius of four miles, when the works would probably have to be moved. The fortunate purchase of the wood on a tract nearby permitted continuation of the work at the same locations near Wilmington but required a longer haul. By November the average haul was five miles, thus increasing the number of teams called for.[37]

The cost of hauling the wood such great distances would soon make the charge of the salt prohibitive. Woodfin, as early as January, 1863, complained that wood was costing two and a half bushels of salt per cord when salt was commanding twenty-two to twenty-five dollars a bushel, and that no one would even consider fifty dollars for wood delivered and that some salt-makers were offering twenty dollars per cord for hauling it short distances. Half his furnaces stood idle while speculators lured away several of his wagoners by prices which it would have been ruinous for him to offer.[38] Touching appeals went out to county commissioners and to Governor Vance, to help provide teams, since Woodfin needed forty teams by the fifteenth of March, 1863. Meanwhile his only recourse was to haul from the woods over his timber road and to throw off the sticks into the field where the bad road began, nearly a mile from the railroad switch. When the railroad on one occasion failed to haul wood because of lack of fuel, Woodfin removed twenty hands from his wood yard and set them under an overseer to chopping wood for the railroad during a period of more than three months, at a time when labor was otherwise unprocurable.[39] The woodyards from which the various works at

Saltville drew their fuel extended from Abingdon to Marion, a distance of thirty miles. Furthermore, the wood had to be cut well in advance of the time when it was needed so as to dry sufficiently to be fit for the furnaces. Some conception of the size of this problem of fuel may be gained by the bare statement that the two furnaces of the Virginia state works operated in 1863 required twenty-five cords of wood a day. The superintendent of those works was forced at times to avail himself of his legislative warrant to impress both wood and provisions. The record shows that he secured no less than 11,550 cords of wood in all by this method.[40]

Woodfin found that many teams and wagons which had been detailed by military authorities never reported, while other teams sickened within a few weeks with remarkable regularity. Rather bitterly, he flung out the charge that men sought service in the salt industry to avoid other service but were determined to perform none. During the cropping season, as he termed the time of harvest, it was particularly difficult to secure teams;[41] and at the Christmas holiday season it proved so impossible to hold the hands that by New Year's Day, 1863, this salt-maker was reduced to eleven wagons with the serious likelihood of having to stop half his kettles.[42]

Worth solved his problem of hauling wood temporarily by building three flat-boats, as we have seen, to carry wood through the sound, so that he was able to reduce the number of teams and hence the price of the salt.[43] After the seizure of his boats by General Whiting he had to fall back on the old method of hauling the wood, but he could not replace the teams he had discharged and he could not have fed them had they been procurable.[44]

As was natural, agents attempted to apply to the solution of this problem the precept, "If the mountain will not go to Mohamet, Mohamet must go to the mountain." Hence, as the wood receded ever farther from the furnaces, some agents conceived the idea of running the brine out to the standing timber. John Milton Worth, as early as September, 1862, reported that he was trying to run the water two and a half miles into the woods. It does not appear that he brought the idea to a happy conclusion, for we know that a year later the haul at those works had reached five miles.[45]

The problem of labor at the salt works presented several aspects. Woodfin experienced some difficulty at the opening of the North Carolina works at Saltville in 1862, because slave owners refused to risk their negroes so near the enemy's lines. These fears were, however, gradually overcome so that he was able to obtain slave labor at twenty dollars a month and expenses for transportation.[46] Guardians in Alabama were authorized by law to hire out the slaves of their wards at the salt works.[47] The danger of having white workers conscripted hung like a sword of Damocles over the head of every salt-maker. Probably Woodfin and Worth were better protected by North Carolina legislation than most producers, but they were not allowed to escape unscathed. For instance, in the early part of 1863 we find the former asking that a dozen teamsters, enrolled as conscripts, be detailed to the works, although under the law, he thought them exempt.[48] In April, 1864, nearly all his coopers were conscribed, whereupon he begged the governor to have them returned, as they were "as important to the work as any other men I have employed. I find it impossible of late to get sacks and therefore have to use barrels exclusively." [49] A more serious situation occurred in December, 1864, when Israel, in charge at Saltville during Woodfin's long absence, had all his men, only twenty in number, conscripted without warning, thus halting the works, as he was left with only three helpers. Israel, who, despite obvious excitement and consternation, was not debarred from prompt action, went to Abingdon to sue out a writ of habeas corpus against the commander of the post, and then sent one of the small group left to Governor Vance for a copy of the contract. To the executive he suggested a legislative act exempting the men certified by the superintendent as absolutely necessary to carry on the works. With unconscious humor Israel wrote the governor, in reporting his action on December 6, 1864, "I feel tolerable salty myself in regard to this matter." [50]

First, last, and all the time the insoluble problem, however, was transportation for the salt to the citizen consumers. Inasmuch as that problem becomes a subject for special consideration in connection with interstate relations, it will be included in this chapter only so far as to state that the impossibility of securing shipment over the railroad was a never-ceasing worry to the agents. As a fitting recognition of Woodfin's zeal and activity, it should be noted that he made a contract with the Nashville and Chattanooga Railroad for carriage of North Carolina salt and secured the allocation of a locomotive, the *Lamar,* together with eight box cars, for the discharge of that particular contract. He secured its renewal the next year, holding the state rarely fortunate, as purchase of a train would be utterly impossible. "I doubt," he added, "if a million will procure a fair train and two cannot be had." [51] The congestion on the railroads led, naturally, to long hauls by wagon. It will be accepted as a matter of course that counties within reasonable distance of salt works would send wagons to bring home the produce, but we find Woodfin planning, as the only possible way to get it distributed to the middle and lower counties of North Carolina, long hauls of a hundred miles or more to points on the railroads beyond the congested section of the railroad. After surveying all the roads out to a good junction point, he became convinced that in the end many of the counties must, as some were already doing, haul by wagon from Saltville through the mountains to a point called Hickory Station, or to Morganton, or to Icard on the railroad line in North Carolina. Though this method might prove expensive, it would be easier and cheaper to haul by wagon all the way to the western counties than to ship partly by rail and partly by wagon. He noted that Ashur, Franklin, Macon, and

Humphrey Counties were hauling distances of 150 to 220 miles. He had hitherto been hiring through cars for the three most distant counties, but no such accommodation could be long expected from the railroad.[52] Governor Vance approved the plan warmly and offered to have a line of wagons established to run regularly during the summer and fall on the lines mapped out. He foresaw also the need of insuring protection from the military along the Tennessee border, as outrages were often perpetrated there on traders.[53]

Woodfin's last report under date of March 21, 1865, shows him busied with plans for this type of conveyance; he advised a train of wagons to run from Max Meadows in Wythe County, Virginia, to High Point and to Hickory. To Max Meadows from Saltville, a distance of 55 miles, he expected to send the salt by flat cars, as that would avoid the worst part of the hauling for the wagons. This plan meant that he was proposing still longer hauls, for the mileage from Max Meadows to High Point is about 350 miles. The counties between the Central Railroad and the Virginia line could probably haul their own supplies cheaper than the state could do it for them.[54]

The superintendent of the Virginia state works, because of the preemption of some of the works exercised by that state, had a further legal entanglement. He became involved in serious litigation, which hampered and delayed the work for months. The Charles Scott furnaces were impressed by act of the state legislature in March, 1864, as will be related later. Although the assessors agreed late in April following upon the amount of the award, the owners secured an injunction and through court delays and appeals kept the state out of possession of the wells until after the middle of August. The company next enjoined the state agents against cutting or hauling wood impressed under the preemption act so that there was further annoyance and hampering of the work. This action effectually stopped the procuring of fuel until the suits were finally dismissed by the courts and the state law sustained.[55]

Minor obstacles, such as dishonesty and misrepresentation on the part of local officials, the troublesome questions of how to furnish salt to refugees in fair proportion to the county's quota, (though the law authorized only sale to the regular county residents), and carping criticisms, despite honest, sincere efforts to serve the cause, had to be accepted as a part of the day's work. Colonel West, as well as his successor, Philips,[56] came under sharp criticism in Mississippi. Woodfin summed up the result of useless hostility to the salt agent very neatly when he said, " * * * if they choose to misrepresent the facts and find fault and newspaper editors choose to malign me and drive off the aid they should insure me we may not expect full success." [57] Worth did not escape, but probably the sharpest and most bitter attacks, as was not unnatural under the circumstances, were directed against Clarkson, who held control, to a very great degree, of the output at Saltville. Witness the complaint of one Robert Logan to the Joint Committee on Salt of the Virginia assembly: "I greatly fear that Virginia has been made to act in such a way toward her sister states as to bring down the judgment of God on the salt enterprise. A gentleman from Saltville states that it is doubtful whether the injury to the country by the Yankee raid at Saltville is greater than the injury resulting from the iniquitous course pursued by the agent of Virginia—The secret of this appears to have been that the agent of Virginia had his individual salt enroute for the South, and used his authority as from the board of public works, to prevent the salt intended for supply from going forward, so that by this means the supply might be diminished and the price increased. If this last be true, is it any wonder that God has sent his curse upon the place?—the legislature owe it to themselves and to the good people of Virginia to purge themselves from the sins of J. N. Clarkson. My own opinion is, that a more stupendous fraud has never been perpetrated than has been effected through his instrumentality; and the wonder to me is, that the president of the board of public works should not see it." [58]

The attacks against him were such that a thorough investigation was made by the Joint Committee on Salt in January, 1865, under direction of the house, especially in relation to his impressment of supplies. In the course of this inquiry Clarkson himself appeared before the committee, and seems to have satisfied that body, for he was continued in his post.[59]

A task which must have irked these commissioners was the preparation of the incessant reports which had to be written under most difficult conditions. Woodfin, for instance, states in his very first report that his cabin was so crowded with county commissioners and men there on business or for instructions that he had to avail himself of the comparative privacy of a portion of the night to get his report written.[60]

The numbers and character of the salt-workers merit brief consideration. Despite the cry of the scarcity of labor at the works, the total numbers rolled up in the various states are impressive. Probably 5,000 men and boys labored on the coast of Florida and found escape from the army.[61] Five to six hundred men is the usual estimate for the coast works of North Carolina at Wilmington, including state and private works.[62] Five thousand is the number usually cited as engaged at the Alabama wells, 1,500 at the Lake Bisteneau region in North Louisiana, while 3,000 for the Steen's Salines in Texas would appear a generous calculation. When the New Iberia mine was at the climax of its operations, it was said to be keeping 400 to 600 men employed day and night. Occasionally bitter complaints were made as when two strong, young, white men were seen driving one wagon, for it was felt that they should be replaced by negroes, and immediately forced into the depleted military ranks.[63] In general the hands at the various salt-works were negroes, men and women, the whites being confined to the administrative positions—overseer, wagon master, superintendent, and clerks. It is interesting that Mississippi considered using her convicts on the work; Philips, general agent, could see no reason why they should not be made serviceable in chopping wood, making barrels, and firing furnaces; he further felt that they could be worked in large enough groups to make it possible to guard them.[64]

## CHAPTER VIII

# LIFE AT THE WORKS

THE CLEAREST picture of the salt-workers and of the life at the works has come down concerning the operations in southern Alabama. With variations to allow for the local setting so that at Saltville the worker is visualized as shut in by the Virginia Mountains; on the Florida coast, as erecting his furnaces on the sands and looking out over the tumbling waves of the Atlantic or of the Gulf; and in Clarke County as moving over a low, swampy bottom which the Tombigbee River claimed at high flood, the scene was much the same. Therefore, it may be permissible to turn the camera on that section of Alabama beside the Tombigbee River, not far north of Mobile, to which thousands of Alabamians and Mississippians thronged during the years of the Civil War.

If one were to visualize the scene, as if from an aeroplane, one would discern the palmetto flats along the winding length of the Tombigbee River and along the Jackson, Steve, and Salt Creeks from near their mouths to a point about three miles up-stream. A little off the main valley of the Tombigbee, extending from about four miles above the city of Jackson near Old Saint Stephens nearly to Peavy's landing, stretched one region of major activity in salt-making—what was known as the Upper Works. This major portion of the works lay on Jackson and Upper Salt Creeks. In a number of places the earth was thrown up as earth-works to protect the salt furnaces from the high water of flood-time. As the eye traveled across the Tombigbee from the south end of this slough, it would be arrested by a vast assemblage of dug wells covering many acres of ground, all giving evidence of having been sunk very recently. A visitor would have been astonished at the magnitude of the works.

In the bird's eye view to which we aspire, we can allow the eye now to sweep to a point six miles south of Jackson, where we would detect a valley lying between an almost perpendicular hill, the so-called Salt Mountain, and a smaller elevation of limestone, known as Saint Stephens. Wells here were scattered all over the valley, some even sunk into the sides of Salt Mountain, so closely did the salt-makers crowd upon each other. This region of activity was called the Central Salt Works.

Another sweep of the eye to a point near Sunflower Bend, five miles further south would reveal the Lower Salt Works, planted from half a mile to a mile from the river, but situated nearer the stream than any of the other wells except a portion of the Upper Works which extended into Washington County.[1]

A visitor would readily grasp the fact that it was not part of the intention of the confederate government to leave these important salt works defenseless against federal gun-boats, which threatened to enter Mobile Bay and take the city which guarded the mouth of the Tombigbee River. Therefore, one should not fail to note Carney's Bluff, a few miles from the Central Works, where there lay planted a fortification to protect the busy salt workers. Just below the Lower Works at Oven Bluff was planted another battery of guns.

A visitor approaching the locality on any of the public roads in 1864 would have found himself in a dense press of wagons choking the roads for miles in every direction before he reached any of the works.[2] Some of the wagons would have been drawn by two mules, many by four, each wagon filled with old syrup kettles, or wash boilers, pans,[3] provisions, poultry,—anything and everything pressed into service to exchange for salt or needed in making it. One driver might hail from Pike County, another from Lowndes County, Alabama, and some even from Georgia and from Mississippi. The wagons would be driven by the planters or their overseers, while fifteen to twenty barefoot negroes plodded along on each side of the wagon. The dust rose in clouds along the roads from the incessant train of mules and oxen, to the great misfortune of the people living along the routes.[4] Large wagons of wood, drawn by as many as six mules or oxen, went creaking on their way to the works. The air vibrated to the sound of axes until it was utterly impossible to distinguish the stroke of one single ax, while the woods at the same time rang with the shouts and songs of the negroes.

When the visiting planter had reached the location where he had decided to make his salt,[5] he would strike camp and set his negroes to work. Some were assigned the task of sinking a well, ten to fifteen feet deep, usually under the direction of the master or the overseer, while other slaves were dispatched with the wagons to the hills about a mile off to haul rock for the furnace. Rocks were easily secured from the region near at hand. Hills only a mile distant from the Upper Works produced an abundance of burrstone rocks, while the hill, known as Saint Stephens just across the creek at the Central Works, was composed of the finest limestone, which was so soft as to permit of quarrying with a crosscut saw. Sawed into blocks twelve by eighteen inches in size, it made a much smoother furnace than the burrstone. The Lower Works also did not lack rocks suitable for furnace building. Some of the very earliest crude furnaces, made in 1861, when salt-makers first began to flock to the wells, were made of sand and rock, stuck together with clay mud, while any old stack pipe served for a chimney.[6] The furnaces constructed thus hastily and for a temporary purpose by the visiting planters were, naturally, rude, as the one objective was to get to the salt-making as quickly as possible. For the remainder of the two or three weeks spent there, the slaves were kept busy chopping and hauling wood to keep the furnace going.[7]

The perfect freedom of the first days of the war soon passed. At that time anyone, resident of the state or neighborhood, felt at liberty to camp anywhere near Salt Mountain and to put up furnaces to make salt for his own use. A certain J. F. Singleton, who owned the greater part of the land where salt was found, raised no objections. Soon, however, two men from Monroe County appeared on the scene with a lease to the land, so that thereafter the transient salt-maker was obliged to get a lease from Dennis and English. The lease usually granted the use of a piece of ground large enough for erection of a furnace and the sinking of a well eight to ten feet square.[8] In 1862 the state legislature regularized the activity by requiring leases of all who wished to make salt for sale.[9]

As the work settled into a regular, if temporary, war industry, the crude devices yielded to better structures. At the Upper Works the furnaces were made of bricks, which were hauled to the wells on a tram-car, running from the works on the Jackson Creek to those on Steve Creek.[10] The small, rude furnaces first stuck up were torn down and replaced by larger ones, thirty by forty feet long, while chimneys towering to seventy feet in the air displaced the earlier pieces of stove-pipe. Wash pots, no longer worthy of a big and important industry, gave place to large, flat-bottomed kettles cast in a foundry, which would hold at one boiling one hundred gallons. To utilize every calorie of heat, several of these kettles graced the front of the great furnaces, while beyond them extended a double row of smaller ones next to the chimney.

After a period of trial of the huge, hundred-gallon kettles, a new departure was made, when it was found that the boiling capacity did not justify the tremendous consumption of fuel, for the question of wood was becoming an ever more serious problem. The new experiment took the form of shallow pans about six feet long and three feet wide, but only about ten inches deep in the center, shelving to five inches at the ends. Five or six of these pans occupied the front of the furnace, holes being drilled in the edges so that they might be bolted together securely. Iron filings and salt placed between the edges of the pans, fusing as a sort of solder, sufficed in a few days' time to make the pans water-proof. Upon this large composite pan was bolted a double bar piece of timber, each three feet wide by twelve feet long, with the result that there had been constructed a tank holding several gallons of water. Next to this queer tank the salt-maker ranged several one-hundred gallon kettles, and then next to the chimney a double row of smaller vessels. The walls of the furnace were usually two feet high, and, if carefully built with proper grates and doors, would distribute its heat from end to end, so that the maker's labors yielded him twenty to thirty-five bushels of salt a day, according to the salinity of the water.[11] To still further reduce the cost of manufacture, some owners of large furnaces built three or four rows of furnaces of the type just described, all to enter into one common chimney. It required considerable skill to build these tall chimneys with a draught sufficient to make them all burn rapidly.[12]

Since the rule soon became recognized that the deeper the well, the saltier the water, the large producers began to bore at the bottom of the dug wells, with the result that a marked improvement in the quality of the product was perceptible. These bored wells were about seven inches in diameter, and were sunk about ninety to one hundred feet into the swamp. Though primitive in construction, they were effective. They were bored with a wooden tube made from some hard wood, in which was placed a valve similar to those often found in buckets for wells around a homestead. The tube was moved up and down by means of a rope attached to a long pole placed on a pivot; the up-and-down motion sank the tube ever deeper into the soft muck, and when it was full, it was drawn up, the valve preventing the mud from falling back into the well. Cypress or some other decay-resisting wood, put together in joints, bored out with an augur, was then used to case the well. The pumps were made of long, green, pine saplings bored with a three-inch augur and made into air-tight joints from ten to fifteen feet long. High scaffolds protected the wells, from which the brine was pumped into a trough which conducted it to the furnace, some fifty to two hundred feet distant.[13]

A large number of private works were to be found at each of the locations. The Central Salt Works were operated by both firms and individuals.[14] At the Upper Works it is certain that hundreds of individuals were making salt for sale,[15] and it is probably true that as many were busy at the Lower Works. The Upper Works were, however, not only larger in area than either of the other two, but many more people congregated there.

The state stored very little salt at the works; as has been indicated earlier, it was shipped, as fast as made, up the Tombigbee to one of the six depots located at various points in Alabama. Naturally, none of the product could move out to other states through the mouth of the river so long as the Yankee fleet hovered around Mobile. The names of some of the boats which ran on the river at that time have been preserved. If a visitor arrived on the day a shipment was being made, he might have seen *The Warrior, The Jeanette,* or *The Old Advance,* moored to the dock.[16] There were, however, many small, private storehouses built of pine-poles. There was no regularity of arrangement for furnaces or storehouses, everyone building as fancy dictated. The state works were, naturally, the best equipped. At the Upper Works Alabama had erected a saw-mill so that most of the houses were built of lumber, though they were, of course, merely rough-frame structures.

The picture presented at any one of the locations was a lively one of great and constant activity. The laborers were slaves, impressed into the service. The different crews were so close together that the grounds looked like one large establishment with numerous employees.

Around the various furnaces negroes were busy day and night,[17] some firing the huge furnaces, others dipping water from the salt kettles in the front row to those nearer the chimney for graining; some were carrying the moist, dripping product to the temporary warehouses, while some, off duty, were lying around the camp-houses resting. Some of the younger negroes might have been seen dancing, clapping their hands, or playing the banjo.

Provisions became scanty at times, but there was no lack of chills and fever in this natural habitat of the mosquito and malaria. Usually, some of the workers were lying in the miserable camps burning with fever. If one may judge from the extensive cemetery on the hill near the Upper Salt Works, mortality among the workers ran high. Small-pox made its appearance several times, but prompt isolation of the cases prevented its spread so that there was no epidemic here as there was of the yellow fever at Wilmington in the fall of 1862. A physician was officially engaged to visit the Upper Works regularly.[18]

Accidents seem to have been of sad frequency. Furnaces were so numerous and so closely packed together that it was not unusual for some hand to make a misstep into a boiling pan. Sometimes such a victim was so badly scalded that he was mercifully permitted to die immediately; other unfortunates lay around the camp suffering for long periods. A story was long recalled of how an old negro, who was dancing too near a furnace, slipped and fell into a boiling kettle to be killed instantly. The heavy fogs which frequently hung over the works because of their low location proved to be the cause of many scaldings.

All phases of life seem to have been reproduced at the works—even to the religious. The Methodist Conference, ever alert to its obligations, during the year 1862 sent a preacher on the circuit of which the Upper Works had been constituted a part. In addition several Baptist preachers are on record as having visited the works to conduct services.[19]

It is conspicuous that the relatively few white workers, almost all in a supervisory post of one sort or another, were either physical weaklings, who had been rejected by the Board of Surgeons, and by the same token were of little use at the salt works, or had been wounded and detailed to the works for "light" service. But it has also been related that occasionally a negro was summoned to write a letter or to balance an account for one of the white overseers.[20]

Perhaps it is not strange that the camps of salt workers should have appealed to deserters from the military service as a haven of refuge. They hoped to escape detection in such an out-of-the-way place as the swamps of southern Alabama, but more likely they hoped, if accepted as salt workers, to be held exempt under the law. So frequently did the deserter bend his steps toward the salt works, that toward the close of the war, every wagon on the way to the works was searched by conscription officers, and not in vain, for a goodly number of deserters were detected and seized from the entering wagons.[21] A great number of refugees from the war-devastated areas also made their way to the salt works, hoping to eke out some kind of an existence until the close of the war. Indeed, the Upper Works presented a cross section of life in the confederacy from preacher to deserter, from maimed army officer to refugee, from skilled superintendent to ignorant black slave, from devoted patriot serving his country by producing salt to that inevitable hanger-on—the gambler.

CHAPTER IX

# VIRGINIA AND INTERSTATE RELATIONS

IN GENERAL, and certainly through the early months of the war, the relation between the several sister states of the confederacy over the salt question was one of friendly cooperation. By the close of 1862 various coast states or states with salines were placing their resources at the disposal of less favored regions or of the central government. Arkansas, for instance, made her salt lands available for the major-general commanding the Trans-Mississippi district, so far as this action did not conflict with her individual interests.[1] The legislative committees of several of the coast states debated the question of offering the courtesy of their shores to each of their sister states of the confederacy for the manufacture of salt.[2] By a coincidence, Florida, at about the same time that Alabama was offering the use of her saline lands, in almost identical language, tendered the use of her extensive coast.[3] Meanwhile Governor Pickens of South Carolina had offered his friendly services to Georgia in procuring a large amount of salt for that state at ten dollars a bushel, an offer which was eagerly accepted by Governor Brown. He ordered 10,000 to 20,000 bushels, as that amount would relieve his people until a supply could be procured from other sources. It does not appear, however, that any portion of the amount was delivered.[4] Alabama, by proclamation of the governor, invited citizens of other states to her salt region, and cheerfully accorded them the right to make salt for their family consumption at any of the works—except those on the state reservations—upon such terms as they could obtain.

To guard against extortion, the export of salt beyond the state was prohibited without the written consent of the governor. The law allowed the governor to lease to non-residents such quantity of the public salt water as he deemed consistent with the wants of the people of the state.[5] The attitude of the salt commissioner of Alabama in May, 1864, would seem to indicate objection to contracts by Mississippi with salt-manufacturers in Alabama for specified amounts of salt, though there had been no objection to the state manufacturing there or to Mississippians making the article for their private consumption.[6]

Virginia, as the repository of the most important salines of the confederacy, was in a delicate position: she must pursue a generous policy with this vital resource toward all her sister states, and yet she must discharge her duty toward her own citizens of assuring them a fair supply. She required more than a million bushels, computing less than forty pounds per capita, whereas fifty per capita had been her pre-war amount. Her first important step was taken in the winter and spring of 1861-62, when a special joint committee of the legislature, created to deal with the problem, considered seriously for months the question of taking over the works from the lessees by purchase or by preëmption.[7] It is apparent that the Stuart, Buchanan Company feared and half expected preëmption, for we find them gravely arguing, in May, 1862, "We are vain enough to think that we can manage the property to better advantage, both for ourselves and the public, than an agent of the state would." They made a point of showing how they had endeavored to secure an equitable distribution of the commodity to consumers at one dollar a bushel on a quota basis of twenty pounds per inhabitant for a year's supply by contracts with the county agents. They furthermore held that their rights in the seven furnaces which they were then operating and in the two under construction ought in all fairness to be allowed them through

1862 and 1863, in which case they would gladly allow the state to put up the erections it wished.[8]

The committee made a provisional contract with the lessees for 400,000 bushels of salt at 75 cents a bushel to be delivered monthly from May 1, 1862, to May 1, 1863, this being the largest amount the lessees could provide because of existing contracts with the central government and with individuals. Though the assembly approved the provisional contract by resolution, it adjourned without passing the appropriation to carry out the contract, thus releasing the lessees from any obligation.[9] Even at the special session called in May, it merely authorized county courts to purchase and distribute salt for the people of their respective counties in the amounts deemed necessary, payment to be provided by county levies or by loans or bonds of the county; the courts might appoint agents or commissioners to carry out the work.[10] Thus it is to be noted that Virginia, preferring to rely on the integrity and honor of the lessees, declined at this time, 1861-62, to purchase or to take over the works.[11]

Governor Letcher, like all the other states executives, found himself obliged to grapple with the problem which the assembly had evaded. By June, 1862, he had sought advice of his attorney-general as to whether he had power to seize the salt-wells and work them on behalf of the people, but the opinion of this official was adverse to such action.[12] The governor's personal visit very soon afterwards to the works so impressed him with the activity of other states in his own state—Georgia, Tennessee, North Carolina had already begun the manufacture of salt, while Alabama was erecting works—that he was spurred to executive action. He was goaded also by the anxiety of Virginia citizens, who were earnestly urging action, by the exorbitant prices of the commodity, and by his sense of obligation toward soldiers' families. He summoned the legislature in special session September 15, 1862, and laid before it the proposition offered him by Stuart,

Buchanan, and Company at his invitation. This would give Virginia the privilege of erecting furnaces and of manufacturing salt at convenient points upon the grounds leased by Stuart, Buchanan and Company to agents appointed by the governor, who were authorized to sell the salt made from the brine furnished by that company at a cost of seventy-five cents a bushel. The state agent could sink wells and pump brine, if the lessees failed to deliver it.[13] The proposition of the lessees at Saltville was unsatisfactory to his thinking, because it exacted of Virginia, in which state the works were located, a larger price for the water than was asked of Georgia, and the same price asked of the other three states.[14]

The legislative body [15] promptly passed an act, authorizing the governor to adopt any measures necessary to provide for the production and distribution of such an amount of salt as he deemed sufficient for the needs of the state. Purchase of the Saltville or Kanawha salt works was, however, explicitly forbidden; but the governor was expressly authorized to seize and control the property of *any person or company* necessary for the production of salt, a provision which threatened, of course, the works erected by the sister states in Virginia. Existing contracts with the confederate states or with any state, or with the Virginia counties, might be disregarded if sufficient salt could not be produced for the people of Virginia, though otherwise such contracts under the act of May 9, 1862, would be respected up to the amount of twenty pounds per inhabitant of the state concerned. The governor was empowered to take control of any railroad or canal in the state in order to transport salt or fuel or other requisite for the production of salt, though interference with government troops or army supplies was forbidden. He was to designate distribution centers for the sale of the salt in the state, to prescribe rules governing the sale, and to fix the prices.[16] Finally, $500,000 was appropriated to carry out the act. Any surplus over the needs of Virginia was to be furnished to the other states, and property seized was to be paid for.

Immediately after the adjournment of the special September session, Governor Letcher again visited the salt works in order to execute the law, which he declared the most perplexing and difficult of the tasks devolving on him, as it required great prudence and judgment in order not to conflict with the central government, with the various states which already had contracts with the salt proprietors, or with the Virginia counties.[17] He found it impracticable to take possession of the works, as the damages for seizure and cost of labor would have absorbed nearly all the appropriation, with no advantage to the people in securing a supply of salt. Furthermore he found the works operating under a system of barter which would have handicapped the state in any operations. He decided, therefore, in the interests of harmony to avoid conflicts by buying the salt. A contract was made with responsible parties for 150,000 bushels beyond existing contracts, at $2.33 1/3 a bushel, this being the utmost they could furnish. A proclamation was then issued on November 15, arranging for the distribution among the counties and fixing the price at three dollars a bushel.[18] In order to afford the widest relief possible, all counties were included in the distribution except those which were in part or in whole under the control of the enemy. An effort to purchase 100,000 bushels of salt at Kanawha failed, because of dilatoriness in executing the governor's orders for wagons before the salines were lost to the confederacy. The governor thereupon secured with great difficulty a contract for 34,000 bushels additional from Saltville, which could not, however be delivered before March.[19] He had already on October 15 issued a proclamation, barring the shipment of salt out of the state, except on prior contracts, until Virginia should be supplied; he also seized some salt of the Georgia Troup Company, which was, however, later released.

The law which was the crux of Virginia's difficulties

with her sister states was that of March 30, 1863. By the time the legislators assembled for the adjourned session in January, 1863, they were thoroughly alert to the situation with regard to the extortion charged against the Stuart, Buchanan Company. They revived the Joint Committee to study the means of providing a supply of salt and to consider amendments to the act of October 1.[20] By the end of March revolutionary legislation had altered the whole relation of Virginia to the proprietors, and, by the same token, to her sister states. The office of superintendent of the state works, carrying for the time a sizeable salary of $5,000, was created, to be filled by the assembly; the superintendent was prohibited from all interest, direct or indirect, in the works. Under the Board of Public Works, serving as a Board of Supervisors, he was placed in supreme charge, to manage and dispose of that portion of the property of Stuart, Buchanan and Company, which consisted of one to two hundred acres, ten furnaces, and equipment, which was leased by the State of Virginia for one year under a contract of March 25 between the legislature and the company.[21] He was empowered to lease real property and to buy any personal property necessary to assure a supply of salt for Virginia; to contract for supplies, to hire all necessary labor, to impress real estate and labor, if mutual agreement proved impossible; and he also was charged with the distribution of the salt. In the event that the original lessees (Stuart, Buchanan Company) failed to comply with their contract, the superintendent was vested with power to impress their property. The law carefully respected the leases of certain furnaces,[22] except that in the event of failure on the part of any one of the lessees to fulfill his contract, the superintendent should relet to other parties or operate the furnace for the benefit of Virginia. The other six furnaces he was directed to lease or to operate for the state according to his best judgment, three of which he did ultimately relet. In order to insure distribution of salt to the counties, the superintendent was empowered to control transportation on the railroads in the state even to the point of impressment. The obligation of distributing the salt according to population, in the proportion of twenty pounds per person, was laid upon him, the salt to be sold at cost in return for cash. Any surplus, after the needs of Virginia had been met, might be sold to the confederate government, or to any of the other states, or to their citizens. A million dollars, appropriated from the state treasury, and the sums resulting from the sale of the salt were to be devoted to executing the purposes of the act. The law provided for estimating the value of any property impressed or of real property used, appeal lying to the circuit court of the county in which the property was located. Monthly reports were required of the superintendent to the Board of Supervisors, and from the latter to the governor; reports to the assembly each session were also stipulated. It is apparent from later conduct that the contract was wrested from the Stuart, Buchanan Company against their desires.[23]

The legislators had become somewhat disturbed over the question whether any allowance in weight should be made for salt delivered wet. The committee learned that it had never been the custom at the Virginia works to make any such allowance. However, another act of the same date sought to raise the quality of the salt by providing for inspection by an official inspector, whose salary was to be paid by a fee of one-half cent per five bushels handled. No inspection was ordered of salt manufactured at Saltville by other states which was not to be sold in Virginia. The following October the committee recommended a slight increase in the fee, as it had not even covered the cost of the materials used in marking.[24]

Although the revolutionary act was passed March 30, it was not until June 7 that Virginia took possession of the works, because of tedious delays in making the assessments for the sum to be paid Stuart, Buchanan, and Company.[25] After possession was secured, John N. Clarkson, the newly-appointed superintendent, found the works in bad condition.[26] Flues and kettles were out of order, and some furnaces had to be rebuilt. Clarkson may justly be considered truly in the saddle by the twelfth of June, whereupon a sum of money was advanced him to purchase teams, wagons, harness, sacks, and other supplies. Since the Stuart, Buchanan Company failed to deliver brine in adequate quantity, the state proposed to buy 40,000 bushels additional for the packing season of 1863-64 from the Buchanan Company and from the Charles Scott Company, deliverable in November and December, 1863.[27]

Impressment of the salt property did occur in behalf of the state of Virginia, but not until the spring of 1864, and then, because of failure to deliver the brine, it occurred by legislative direction and not by action of the superintendent. To be sure, Clarkson, irritated and hampered in his work by the original lessees, had suggested impressment of the Preston well. The company had, undoubtedly, interfered with the efforts of the superintendent to hire teams which had been working for them. In one such instance, when he reluctantly resorted to impressment, he found the lessees urging the parties whose teams he was impressing to litigate his right to do so and promising to "stand by" with their financial aid. They even silenced counsel who had been retained sometime earlier by the superintendent. The company had also offended the committee during the negotiations for renewal of the lease by greedy demands for transportation of excessive amounts of wood, (sixty-six cords a day), by some misrepresentations in regard to the lease of one of the furnaces, and by stipulating the exemption of *all* their property from liability to impressment. The ultimate result was that the committee recommended impressment of the Preston wells at "the first recurrence of failure to deliver brine." [28] But the legislature was ready to act immediately. A law was, accordingly, passed on March 8, 1864, which directed the superintendent to impress until June 8, 1865, three double furnaces, called the Charles Scott furnaces, which were nearly new, whereas the six furnaces leased and taken over the year before had been worn out and dilapidated. He was also directed to seize the Preston well with all lands, equipment, and appurtenances necessary for the proper working of the furnaces leased by the state, together with all the blocking water furnaces. In case of the failure of the supply of brine, the superintendent might take possession of the other wells in order to supply the state furnaces to their full capacity, permitting only the surplus beyond state needs to flow to other furnaces. The Board of Supervisors was given plenary power in order to procure transportation for salt or wood to hire engines and cars from other roads and to place them on the Virginia and Tennessee Railroad. Due regard was had for the needs of the army by providing that the Board of Supervisors should supply salt to the central government on such terms as might be arranged with the secretary of war. The usual provisions were made for assessing compensation and allowing appeal. To provide the requisite capital for this large enterprise, the state appropriated $2,000,000.[29] It should be noted that these preemptions still left the Stuart, Buchanan Company some wells and the means of continuing business.

It was the opinion of Governor Letcher in the fall of 1864 that the interests of all parties would be best served if Virginia were to undertake to supply salt to the other states, in return for the supplies sent in by their state trains.[30]

The friction between Virginia and the other states arose over the question of transportation of the salt. Railroad transportation was a serious problem in the confederacy everywhere and always,[31] but it aroused emotions to a high pitch in this particular controversy.

It must be understood, in the first place, that the movement of salt presented difficulties elsewhere than in Virginia. The confederate government had, of course, priority over everything for the transport of troops and army supplies. During the movement of troops to and from Vicksburg, extending from October, 1862, to June, 1863, freight of every description was held up; cars already loaded were ordered unloaded by the military authorities.[32] The experience of Governor Pettus in getting salt from the Louisiana mine moved from Vicksburg in the late fall of 1862 is a case in point. He wrote an official of the Southern Railroad that the people were starving for salt, and yet they must express it or suffer long delays if they ventured to ship by freight. The reply stated that under the orders of General Smith private freight could be shipped from Vicksburg only one day a week, the rolling stock being devoted the rest of the week to government sugar, and that the congestion was so great that it would take four to five months to move out the government and private freight accumulated at that point.[33] Despite the concession of General Pemberton that at Vicksburg one car might be loaded daily exclusively with salt for the suffering families of Mississippi, the cars continued to be preëmpted for the removal of troops, with the result that the salt was delayed from day to day until by the middle of January citizens were still complaining. The outburst of a soldier who had purchased salt and sugar in Vicksburg for the suffering families of men in his company out of his paltry soldiers' pay, eleven dollars a month, and who had found it undelivered a month later was probably futile, but it arouses a certain sympathy even now. "In the name of God, I ask is this to be tolerated? Is this war to be carried on and the Government to be upheld at the expense of the starvation of the women and children of the country? I hope not." [34] Alabama had similar experience with some salt halted at Memphis [35] and at other places. On February 13, 1863, a county agent wired Governor Shorter that salt shipped by Captain Speed from Saltville had reached Montgomery, but that he could not get it past some burned bridges; on August 17 the agent for Lawrence County complained that he had received only 200 sacks of his quota, while somewhere between Saltville and Chattanooga 200 more sacks were being delayed.[36] A North Carolina agent met the same delay in getting some corn through to the salt-works at Wilmington. The problem did not improve, unfortunately, with the progress of the war.

Even the confederate government and army officers had to grapple with the situation created by a scarcity of rail transportation for their salt supply. General Floyd in the first year of the war found himself shut off by the enemy from the salt supply at Kanawha and Bulltown (West Virginia); but he attempted to procure some for Lee's army from a small salt works on the border of Kentucky by a train of wagons.[37] An enterprising colonel, operating in Virginia in February, 1862, determined to make the experiment of transporting provisions from Peterstown to Rock's Ferry by canoe, following the example of the people of that neighborhood in bringing their salt from Mercer Salt Wells in that fashion. Two canoes, each forty feet in length and bearing 2,000 to 2,500 pounds, were prepared, and should, it was estimated, make the round trip in less than two days.[38] General Taylor was concerned in the fall of 1862 with keeping the Atchafalaya open, since it afforded the only ready means of transportation for salt from the New Iberia mine to the Red and Mississippi Rivers and, hence, to the portion of the confederacy east of the Mississippi.[39] At the close of that year the commissary-general was explaining to President Davis that salt had "to take its chances for transportation over long distances, upon uncertain roads discordantly connected."[40] The government disclaimed any power of control over the railroads other than that of preference in the transportation of munitions and supplies,[41] though it did create the office of superintendent of railroads.

The first fear and distrust of Virginia on the part of her sister states was voiced in October, 1862, immediately after the legislature of that state had reposed in the governor power to seize the property of any person or company. It was felt a grievance by Georgia, Tennessee, and North Carolina that Virginia, after adjourning her extra session of April, 1862, without action on this important question and so encouraging the other states to erect their works at Saltville, threatened seizure just as they were beginning to operate successfully. The action brought forth a spirited challenge from Governor Brown of Georgia, in which he appealed to Virginia to act with the good faith consistent with the character of the "Mother of States," and declared his readiness to carry out any instruction of the Georgia assembly "for the defense of the rights of the State of Georgia to the last extremity."[42]

As early as the fall of 1862 the clamor from the states which had set up works in Virginia over the impossibility of shipping their salt out was unanimous and strident. Governor Shorter carried his complaint straight to Secretary of War Randolph, charging that only about one-third of the salt made daily was taken away and that the portion shipped was for Virginia use, while other states were denied all facilities.[43] Woodfin, salt agent for North Carolina, had not much more than begun operations when he voiced fear over the problem of transportation.[44] Georgia also appealed to the secretary of war and received that official's assurance that transit of salt should be facilitated.[45] The problem did not improve as the war developed, for salt purchased by Georgians at Saltville in October, 1863, still lay at the station unshipped the next March.

Complaints were lodged against the railroads themselves of inefficiency and partiality toward the private companies, especially Stuart, Buchanan and Company. The Alabama agent at Saltville charged that the Virginia and Tennessee road shipped eight carloads daily for that firm, while not one sack had gone out for Georgia or Alabama in two weeks. The worst aspect of the situation was that he saw no prospect of improvement, indeed, only the reverse as the bad weather of winter set in, when land-slides frequently blocked the branch road for weeks at a time.[46] The means of transportation on the branch road were utterly inadequate; the road carried out only 100 sacks per car, about 2,400 bushels per day, while the works produced 7,200 bushels or three times that amount daily. The salt was accumulating in the sheds of the firm manufacturing salt for Alabama so as to inconvenience and hamper their work. The agent's appeal to the local railroad officials, stating the pressing necessity of the Alabamians, was answered with the plea of instructions from the railroad president to transport eight car-loads a day for the local lessees and of the inability of the locomotives to carry more on the heavy grade of the branch road.[47] Governor Vance [48] and Salt Commissioner Worth were constantly pressing on the railroad officials their dependence on the roads for transportation of salt to the western counties of North Carolina. In January, 1865, General Lee denounced the wretched management of the Piedmont Railroad, showing that salt and corn in "immense quantities" had been left piled in mud and water while the army was starving.[49]

One county agent from North Carolina even went so far as to accuse officials of the road of a species of graft; he declared that in their cupidity they would neither allow trains from other roads to pass over their tracks nor transport the salt "unless you cram their pockets with the Almighty dollar," and that the citizens of Cherokee County were not able to gratify their demands.[50]

Confronted with the problem of transportation, the states naturally addressed themselves to finding a solution. Many of the state and county agents tried the expedient of buying an additional train, but cars were not to be had. In North Carolina Governor Vance found that the entire locomotive power of one of the railroads from Saltville to North Carolina was controlled by the confederate government, while he could not send engines from the other road because of a different gauge.[51] Woodfin pressed the matter so far as to try to get a locomotive from the government. At the Tredegar works in Richmond a man said that by widening the flange of the wheel it might be possible to run one narrow-gauge locomotive on the broad-gauge road at ten miles an hour, but at any greater rate of speed it would not keep its hold on the track.[52]

Georgia fared somewhat better. With power to command rolling stock because of state ownership of a broad-gauge railroad line in the state, she promptly [53] proposed to send trains of her own to bring the salt out of Saltville.[54] In this course she was undeterred by the fact that nearly one-third of the rolling stock of the state-road had already been seized by General Bragg during his campaign in East Tennessee and that several engines and nearly 200 cars had been lost while on such service. A firm executive inspired firmness in the legislature, and so Governor Brown was directed before the close of November, 1862, to send trains from the state road and, if necessary, to impress cars and engines of the different railroads operating in Georgia, in order to send for the state salt. Such action, however, was not to interfere with the requisitions of the confederate military authorities.[55] The officials of the other roads of the state, confronted with the danger of impressment, responded to the governor's appeal, and provided three trains. The speed with which things moved to secure transportation for the salt under such a threat of impressment is amusing. Telegrams flew back and forth on November 22, showing that the Southwestern Road could spare one full train of fourteen cars and would provide an engine and six cars toward a second train, if another road would complete the train.[56] A few days later, the assembly, spurred on in its good work, requested the governor to have salt belonging to citizens of the state brought to Georgia from the various points of manufacture, and vested him with power to impress engines and cars for that purpose.[57] Meanwhile Governor Brown had secured from the various railroads of the state their assent to carry all salt imported by the state free of charge to the depots of deposit and thence to the depot of distribution. He, in turn, had ordered the state salt, that made at Saltville by the two Georgia-sanctioned private companies, and all salt purchased at the works for personal use or by county associations to be carried over the state roads gratis. It is almost amusing and certainly typical of the condition of the railroads that on June 17, only a few months after Governor Brown's energetic efforts had placed a train in Virginia, he was notified that the engine sent was in a bad condition and in the shops for repairs, so that he had to have it replaced.[58]

A further complication in the problem of trains of one state running on the tracks of another was the very high tolls exacted of the visiting trains. The Eastern Tennessee and Georgia Railroad, for instance, charged Georgia full freight, though it disclaimed any responsibility for loss or damage.[59] A record for November, 1863, indicates a charge of $900 for each load of salt run over the tracks of the Virginia and Tennessee Railroad, exacted by the directors of that road, whereas the regular freight was only $180. On such a basis it was computed that merely to run a train of ten cars over the tracks would cost $7,200![60] In the course of the year 1864 North Carolina hired a train from some Virginia road at $2,000 a day, and had to pay in addition for the use of the tracks, for fuel and services, an amount which constituted seven-eighths of the regular freight. Hence, an official calculated that to send one train of eight cars, carrying 2,400 bushels of salt from Saltville to Danville, North Carolina, and back, a trip which would require twelve days, would cost $24,000 or ten dollars per bushel for freight.[61]

Though the president of the South-western Road of Georgia dispatched a train again on December 11, it failed to take care of the rapidly accumulating salt. Governor Brown's first thought was to furnish sacks and provide a rough store-house at Saltville, purchased, rented, or built for the purpose, but when this project proved impossible, he suggested lending the salt to Stuart, Buchanan Company, with the written understanding that it should be repaid as soon as Georgia could secure transportation.[62] Hence, he proposed in April, 1863, to set aside a special train from the state road to be used solely to transport the salt to Georgia and to carry needful supplies to the furnaces.[63] This proposed solution of the problem, however, miscarried, for the Virginia and Tennessee Railroad now refused to allow Georgia trains to run over its tracks. This, of course, implied no discrimination against this one state, for the superintendent of the Virginia road had by March, 1863, determined on a general prohibition against all "foreign trains." The superintendent held such a procedure injudicious as well as dangerous and embarrassing to the operations on the branch road from Bristol to Saltville, since the employees of the foreign roads would be unfamiliar with the peculiarities of the Virginia road, with its regulations, and would not be responsible to the officers of the road.[64] It would obviously help to wear out the tracks and injure the road-bed, especially if large, heavy engines were used; furthermore, it could greatly impede the domestic trains and thus deprive Virginia of her own means of transportation.[65] The superintendent hoped to meet the situation by establishing a daily service between Saltville and Bristol.[66]

Governor Brown concluded, therefore, to accept the proposition of the superintendent of the Virginia railroad to hire out two of the Georgia trains to that road for the purpose of supplying all the Georgia works with wood and of bringing Georgia salt out from Saltville to Bristol, despite the fact that the compensation offered was much less than the actual value of the hire of the train.[67] A warehouse at Bristol for the salt then became a necessity. The state thus faced frankly dependence on the East Tennessee and Virginia and on the East Tennessee and Georgia Railroads for transportation of the salt from

Bristol to Dalton, Georgia. When, late in May, an engine and train of eight cars was sent from the state road for this service, it was expressly stipulated that while the train would be under the control of the Virginia and Tennessee Railroad, its use was to be devoted to the transportation of salt to Bristol and of wood and supplies to Saltville for Georgia and Georgians in a stipulated order of preference.[68]

The plan of a conference of the executives of the principal railroads with a view to solving the problem of transportation in connection with salt was worthy of greater success than that with which it met, but little could be achieved in the face of war conditions. Called at the suggestion of the assistant adjutant-general at Richmond, the railroad convention, consisting of delegates from the various railroads of the confederacy and of representatives of the states, assembled at Augusta, Georgia, on April 28, 1863, and considered the best means of transporting supplies to Virginia, and of expediting the movement of salt from that state. Among other measures it suggested furnishing to the Eastern Tennessee and Virginia Railroad sufficient iron to construct four miles of track to enable it to supply the works with wood. The suggestion seems to have reached the governors and the secretary of war, but, unfortunately, never to have been translated into action.[69]

Meanwhile, when direct efforts with the Virginia railroads failed, the state executives and private individuals did not hesitate to bring charges directly to the head of the Virginia government, a procedure which was productive of great friction between the states. A North Carolina citizen stigmatized the Virginia policy to Governor Letcher's face as "illiberal," for the former felt that a citizen of North Carolina was as much entitled to the privilege of transportation in Virginia as a citizen of that state. Governor Letcher resented this reflection, and in a letter to Governor Vance pointed out that Virginia had protected all the contracts made for the manufacture of salt at her works at a time when she herself was suffering for that article.[70]

Salt for Virginia had also accumulated at the works—to a greater extent than for some of the other states, according to Governor Letcher's contention at the close of the year 1862—so that he had been driven to the necessity of procuring trains from other roads than the Eastern Tennessee in order to move the state salt out to Virginians, who were suffering equally with the citizens of the other states. "No one desired more ardently than myself," he declared, "to secure amity and harmony between the states composing our Confederacy, and all familiar with my course will bear me witness that my action has been in strict accordance with my professions. To be successful in this struggle, I have felt that we must be united, and I have been particular to avoid everything that was calculated to produce discussion and divisions, and have spared no honorable effort to cultivate and encourage a spirit of comity and harmony." [71] The feeling had been growing, however, that Virginia was claiming all the transportation after the confederate needs had been supplied, and the question was being agitated whether the confederate government should not take possession of the railroad so as to equalize the transportation among the several states.[72]

Feeling reached a climax when the Board of Public Works in Virginia adopted certain regulations in regard to transportation on the railroads of the state. It must be recalled that in the preceding March, Virginia had changed her policy, and was now actively engaged in manufacturing salt. In order to adjust the question of transportation, the Board on August 31, 1863, established the regulations which conceded priority on all trains to the confederate government, and then claimed for Virginia the use of trains owned by other states after they had removed all salt manufactured by those states, thus asserting her right above private individuals or corporations. A few days after the regulations had been announced, the salt agents of Georgia and of Alabama presented themselves before the board in Richmond and received a full explanation, with which they expressed their entire satisfaction.[73]

A few weeks later Governor Brown sent Colonel B. H. Bigham, who, besides being head of one of the salt companies at Saltville, was a member of the Georgia assembly, to Richmond to lay the difficulty of transportation before the governor and legislature of Virginia.[74] The Virginia legislature, after mature consideration of the complaint, established by legal enactment on October 31, 1863, an order of priority in regard to the transportation of salt: first, the confederate government; second, the states owning or employing such trains, the certificate of the governor of the state in question being made conclusive as to the character of the works employed in the manufacture of salt; third, the state of Virginia; and fourth, private individuals or companies of the states owning the trains.[75]

In order to help move the salt out from Saltville the Virginia legislature made an agreement with Stuart, Buchanan and Company, by which that firm became entitled to the exclusive use for November and December, 1863, of all foreign trains which they had or might introduce on the Virginia or Tennessee road. Five foreign trains were accordingly used by the firm with the result that the movement of the regular trains of the road became greatly impeded and with the further result that by the first of January there had accumulated at Saltville 100,000 bushels of salt and that there was such scarcity of wood that several of the furnaces were idle.[76] The suspicion was perhaps not unnatural that the company was using the trains for the transportation of their private salt.

After Virginia impressed a number of furnaces in March, 1864, and undertook to supply the confederate government with salt for the army, she determined upon what she considered a "just and equitable arrangement" to apportion the transportation of the army salt equally among all the states engaged in manufacturing salt in Virginia. At a meeting of the Board of Public Works held at Saltville in March, 1864, the agents of the sister states concerned appeared before the board to remonstrate against the rule. The board finally concluded to require of foreign trains transportation to Lynchburg of only one load per month, which was but one-fourth of their capacity, less than the requirement under the law, and less than the amount constantly required of Virginia trains. Although the agents professed satisfaction with the arrangement, the foreign trains failed to comply with the requirements, whereupon the board directed the president of the Virginia and Tennessee Railroad to forbid the visiting trains the use of the road unless they transported their arrearages. When the North Carolina agent declined to haul the government salt in November, he found his train placed on a side-track.[77] A little later the visiting trains were required also to haul wood for the furnaces making confederate salt. The agents of the four states concerned drew up a protest under date of November 30, 1864, in which in some of the strongest language to be found in this heated controversy they condemned the course of Virginia, sending copies of the protest to their respective governors and to the chief executive of Virginia.

"Therefore now," they claim, "since it is apparent that the production of salt at that place has greatly decreased under the operation of Va. salt laws and the interference of the Va. Salt Agents, it is preposterously absurd for it to be said that this road with its short western connection is unable to do the diminished hauling left to be accomplished. * * * We scarcely know to whom we should appeal to rescue our States from this unwarrantable appropriation of their transportation—Again—Our contracts were made with the proprietors of the works long before said works were rented or impressed by the State of Va. Have we no rights under these contracts? The States of Georgia and Tenn. have a joint train engaged from the East Tenn & Va R.R. to be used on this road—Since learning of this unreasonable demand the Superintendent of said road peremptorily refused to permit his trains to come in & be subjected to such interferences and such other terms as the will of these gentlemen may suggest to them * * *"

"Can it be that under the Saintly pretense of keeping down high prices transportation is to be monopolized and delays thrown upon the other States while Va Salt occupies their markets. Can it be that under a report adopted by the Genl. Assembly in which it is declared that Va. claims no priority in the use of foreign trains, an attempt is made to justify this obnoxious tax in kind upon the transportation of her sister States—Can it be that under the idea of making cheap salt, such restrictions are to be thrown around us as will compel us to abandon our works—that Va may engros the sources and the channels of the salt supply and thus build up one grand overwhelming monopoly—unrestrained by competition." [78]

The agents immediately followed up their protest with a personal visit to the Board, requesting suspension of the requirements until replies could be received from their respective governors.[79] These reports of the agents begot, in turn, indignant excitement in their home states. The North Carolina legislators, interpreting the statement of their agent as meaning that their trains, hired for the transportation of North Carolina salt from Saltville to Danville, had been diverted to the use of Virginia, condemned the action as "a serious departure from the courtesy of states" and directed the governor to forbid Virginia the use of North Carolina railroads unless reparations were made for the injury done.[80] Governor Vance, interpreting his behest literally, forbade exportation of articles to Virginia over the state railroads, not, he declared, "by way of retaliation,[81] but as a precautionary measure rendered necessary by the deficiency in the supply of salt which would be thus produced." A long letter, part explanation, part defense, followed from Governor Smith to Governor Vance, the crux of which was the statement of the cause of the order: the conviction that private cupidity might interfere with the supply of the states, and prevent cheap salt from reaching the citizens. He pointed out that nearly the entire transportation of salt for Virginia passed over parts of that one railroad, which, in addition to being burdened with vast quantities of salt, tapped a fruitful region, abounding in iron, plaster, grain, and hay, so that it was impossible for one road to do the entire work of transportation.[82]

The pathetic side of this controversy and of the congestion which provoked it is emphasized by Governor Smith's calculation that a single train of cars per day would carry more wood to Saltville than the works would consume, while a single train out, carrying 3,000 bushels, would provide one-third more transportation than the whole salt supply of the confederate government and the state of Virginia required. But Virginia could not produce that single train of cars for that purpose.[83]

A further point at issue between the two states, serves to show the complexities which can surround a very humble and simple subject. Governor Vance alleged that considerable quantities of salt had been sold in North Carolina by a Mr. Gilchrist, represented as an agent of Colonel Clarkson, the superintendent of the Virginia State Works. He drew from that fact the inference that the Virginia roads had the capacity to transport more salt from Saltville than was required for her citizens or for the confederate government. The explanation is curious. When Virginia acquired by contract certain furnaces of Stuart, Buchanan and Company, it assumed their contract with Colonel Clarkson. Under this contract they hired from him 121 negro men at ten bushels of salt each per month, and twelve negro women at five bushels per month each—making an aggregate of nearly 1,300 bushels per month, which necessarily must find a market or be useless to the owner.[84] The total amount of 10,800 bushels per annum was thus transported on the state roads for his benefit under a special act of the assembly. Gilchrist was Clarkson's agent for the sale of this salt and Virginia agent for the purchase of supplies of food and other articles necessary for the operation of the salt works.[85] The explanation further pointed out that the barter of salt was imperatively necessary, as the only means of procuring supplies, since salt possessed a greater purchasing power than money.

A report by the Joint Committee on Salt of Virginia in relation to the whole controversy, setting forth at length all the circumstances, was approved by the assembly and ordered sent to the governor and legislature of North Carolina.[86] Thereupon Governor Vance made *amende honorable,* admitting the justice of Virginia's position and expressing regret for the hasty condemnation of North Carolina. He agreed to revoke his order against the transportation of Virginia goods in his state, but put a certain edge on the perfect equity of his action by declaring that he would thereafter "require trains from your state to do only the amount of transportation here as is imposed upon ours in Virginia." Undoubtedly a part of the difficulty, as both governors agreed in saying, lay in the abuses which had crept in, as many thousands of bushels of salt were shipped as state salt, which really were the property of private individuals, who sold it in North Carolina markets at seventy to eighty dollars a bushel, while thousands of bushels of state salt lay useless at Saltville.[87]

Therewith the question might seem to have been ended, but it was recommitted to the Joint Committee on Salt in Virginia, and the North Carolina agent, Woodfin, appeared before that body on February 15 and again on the 17 of the month, 1865.[89] The legislature adjourned on February 18 without final action on many matters. With the controversy in this situation, the close of the war put an end to the entire question.

## CHAPTER X

# CONTRABAND TRADE IN SALT

IT REQUIRED no great perspicacity to realize that salt would sooner or later be declared contraband of war; the only question was how soon it would be so recognized by the federal authorities. The first clear note in this direction which the writer has observed was sounded by Sherman in July, 1862, when he denounced the trade in salt and cotton and declared salt as much contraband of war as powder. He could not, he held, permit it to pass into the interior until a district was declared open to trade. In his trenchant phraseology he declared, "If we permit money and salt to go into the interior it will not take long for Bragg and Van Dorn to supply their armies with all they need to move. Without money—gold, silver, and Treasury notes—they cannot get arms and ammunition of the English colonies; and without salt they cannot make bacon and salt beef. We cannot carry on war and trade with a people at the same time." [1] Still pressing the same point a few weeks later, he wrote Halleck in August of a steamer which he had seized for exchanging salt for cotton without military or custom-house permit. He couched the justification for his action in the following words, "Salt is eminently contraband, because [of] its use in curing meats, without which armies cannot be subsisted." [2] That the civilian population so recognized it is apparent from the fact that that same officer declared that he was finding salt concealed in every imaginable shape and form.[3] Despite Sherman's earlier action of August, he expressed some doubt as late as October, 1863, about "salt for curing meats, medicines for curing wounds and sickness," and suggested that Admiral Porter might resolve all doubt by preparing a list of contraband which the former felt certain would be accepted and published by Secretary Chase.[4] Meanwhile, though it does not appear that Admiral Porter ever submitted such a list, action was taken by the government, for stringent orders from Washington had reached Butler and Farragut by the middle of December, 1862, prohibiting the shipment of salt and other supplies to the enemy's lines.[5]

The border lines where the two armies came in contact with each other and the even more extended line dividing federacy from confederacy, stretching for hundreds of miles, afforded just the same facilities for trade in salt that they did for trade in all other articles of contraband. That such traffic was an active and lucrative profession the host of shady illicit traders, following in the wake of the army and gravitating back and forth across the lines, attest. That the United Sates fully appreciated the importance of keeping salt from the confederacy is clear.

Obviously, the smuggling of so bulky an article as salt through the lines from the United States could not yield enormous amounts, except possibly in the aggregate;[6] just as obviously we cannot hope for exact data or definite figures. But it is our duty to try to piece together the scraps of information which the records yield. We find from Sherman's pen a hint that a large supply of salt went south from Cincinnati, for he declared that salt enough to supply an entire army for six months went from that point, and this at a time when salt and cartridges were the two great needs of the confederate army. More positive for this early period is Grant's report of December, 1861, to General Buel that a large trade in salt, powder, caps, and domestics was being carried on in the vicinity of Saint Louis.[7]

By the middle of the second year of the war, the evidence of a large trade in the commodity in the Mississippi Valley region is indisputable. An officer commanding at Cumberland Gap, Tennessee, in August, 1862, captured eight barrels of salt, along with a small lot of swords and rifles, which had been smuggled through the mountains.[8] General Lew Wallace established the fact that there was a scheme to smuggle salt, supplies, and clothing through the lines in Tennessee to the rebel army, but expressed doubt as to his ability to stop it.[9] Sherman charged that all the boards of trade above Memphis were shipping salt south, as the demand for this article to make bacon was so great that many traders succeeded in exchanging loads of salt for cotton. He, accordingly, addressed a letter to the board of trade at Memphis with the object of securing better control of this trade.[10] In a complaint to Secretary of the Treasury Chase he declared it impossible to keep salt, powder, and lead from reaching the south.[11] Rosecrans had to contend with a brisk trade in salt and other contraband goods going on from Pittsburgh in the Mississippi area at the same time, for which he felt that the commanding officers should be held responsible. No fewer than twelve barrels of the commodity had been seized *in transitu*. He regarded patrols as necessary to prevent unprincipled sutler's clerks and discharged soldiers from selling whatever they pleased of contraband.[12] The same complaint of large quantities of salt, flour, coffee, leather, and sugar moving south from Franklin, Tennessee, without a permit was reiterated to General Rosecrans in December.[13]

The stream of trade seems to have grown wider in 1863 rather than to have diminished. The reader is, however, forced to wonder how three hundred barrels of salt could be smuggled out of Hickman (evidently Tennessee), during two days, as an army report states was true in August, 1863, which statement also asserts that smuggling from there was constantly going on.[14] From North Carolina comes a statement supported with considerable evidence that many of the inhabitants near Newbern had abused their protection papers by smuggling out salt in larger quantities than was needed for home consumption.[15]

The scraps of evidence of illicit trade in salt for 1864 become fewer, though we must realize, of course, that this fact may not necessarily mean that the federal officials had grown more effective, but that the traders had grown more adept in smuggling. However, some of the scraps are very interesting. A certain report from Missouri in October, 1864, indicates the unearthing by a scout of five barrels of salt buried in the sand at the foot of Wolf Island, in southeastern Missouri, and supposed to have been hidden there with a view to smuggling it into Kentucky.[16]

Equally telling is the evidence of illicit trade from the confederate side. Governor Milton's testimony on the subject, coming from a civilian, is striking and indicative of the degree to which this trade had become a profession. Writing to the representatives of his state in the confederate congress in August, 1862, he says: "Since then other vessels have left our ports with cotton, some of which have returned with coffee, salt, and other articles of merchandise, which the owners or their agents have disposed of at the most exorbitant prices to the citizens of this and adjacent states. Some of the goods were manufactured in the United States, and over the manufacturers' stamps upon the goods the names of English manufacturers were stamped, which upon being removed exhibited the cunning device of Yankee villainy, thus confirming a suspicion which I had entertained, that frauds were perpetrated under the pretense of loyalty to the South." [17] He further expressed his conviction that this "nefarious and profitable traffic" was being carried on under false professions of friendship and loyalty to the south. He believed that individuals living in New York, Boston, Havana, and Nassau, had placed partners, cloaked as confederates, in various cities of the confederacy. A three-cornered trade, he declared, harking back to the evasion of navigation laws in colonial days, was again in existence; agents in Havana or Nassau received the merchandise sent from northern cities and the cotton shipped from southern ports, an exchange of commodities being thus effected.[18]

The evidence from an officer in the confederate army, writing of the conditions in Mississippi at the close of 1863, is very striking and important. The people of De Soto County (Mississippi) "have been abandoned by our army and left open to the raids of the enemy since the fall of Memphis in the summer of 1862. * * * * They were cut off in the middle of summer without having made preparation for such an event. They could not at once make blankets, shoes and clothing; * * * * and above all they could not make or obtain salt, without which they could not live and even if they could have purchased salt in the Confederacy, the railroads were occupied by the army and they could get no transportation for it. Under these circumstances they traded with the enemy, and the husbands, sons, and fathers in our army of the women in North Mississippi were supplied with many articles of clothing and comfort that came from the enemy's lines. Salt was obtained from the same source, and almost every pound of meat that our army consumed from March until Vicksburg fell in July, was obtained from Northern Mississippi, and was cured by salt bought from the enemy, * * * and when these orders have been enforced against old men and women who have trudged day and night through the mud to obtain a little salt, our soldiers have almost revolted at it." [19] He goes on to tell a pathetic story of this illicit barter. A poor soldier's wife, with several small children and an aged father to support, had struggled until she had sufficient means to buy one bale of cotton, which she took to De Soto County, to exchange for salt and a few articles of family use. She was caught at Tallahatchie on her return, and notwithstanding her heartbroken grief, her goods, little truck cart, and twenty cents were ordered confiscated.[20]

From various officials, especially the governors, came distinct encouragement to confederate citizens to court trade with the enemy, though it must be added that the central officials at Richmond and army officers were reluctant to assist the United States in procuring cotton, however great the need of their own citizens for the salt and medicines for which it was exchanged.[21] Governor Pettus contracted with a private citizen for one thousand sacks of salt, which he ordered to enter the lines unmolested.[22] Many applications of citizens for permission to take cotton through the lines under the promise to bring out clothes, salt, firearms, and ammunition are to be found.[23]

The executives of North Carolina did not cavil to present to the citizens of their state and to the confederate authorities the advantages of this trade. Governor Clark admitted to a military officer during the summer of 1862 that he had approved of North Carolinians purchasing salt, which he understood to be available in Plymouth, near Albemarle Sound. He disclaimed, however, recognition of the fact that in order to produce the salt at Plymouth citizens would be obliged to pass in and out of the enemy's lines, for at that time, in fact, the foe had not garrisoned Plymouth nor had the military lines been drawn.[24] Governor Vance followed in the same path of sympathy with such trade. After he had received applications from several citizens of North Carolina for permission to obtain a supply of salt from within the enemy's lines, in exchange for cotton and turpentine, he directed in October, 1862, a tentative inquiry to Richmond, pointing out that his people were suffering for salt and that their supply had been greatly diminished because of the outbreak of yellow fever at Wilmington. He pointed out the danger, in the event that the privilege of such trade were denied to reputable persons, who could be trusted, that an illicit trade might be inaugurated which would injure the cause and lead to desertions. He ventured the view that no public harm could result if the privileges were confined to discreet persons who should be subject to the supervision of the military authorities.[25] Whether the response of the Richmond authorities to his proposal was cordial or not does not appear, but it is certain that he pursued a line of action in accord with his recommendation, for scarcely a month later he indorsed on a request for permission to go to Plymouth to buy salt, the words, "get as much salt as you can without compromising the Govt." [26] By the following February, the policy at Richmond seems to have crystallized in the direction of sympathy with Vance's ideas, for a confederate general wrote him early in that month that military orders were to encourage loyal citizens to bring produce, beef, pork, salt, and other necessary articles for personal use from federal territory over the Chowan River. The general was to lend them aid, escorts, and every possible assistance, though he would not permit the carrying over of cotton, tobacco, or other contraband for exchange. However, in the next breath he implies that he would wink at "illegal traffic in order to get things absolutely necessary." [27]

From Virginia also come hints late in the war, of this trade with the United States. On one occasion, when the County Court of Washington County, Virginia, found itself with a surplus of 680 bushels of salt, far more valuable than confederate currency or scrip, the court directed it carried to Tennessee, which was then, in 1864, within the Union lines, to be exchanged for corn to relieve the suffering families of soldiers.[28] On the very eve of the collapse of the confederacy, March 27, 1865, we find Governor Brown, addressing the "Federal General Commanding Savannah" directly to learn the terms on which the United States government would "exchange salt, iron, agricultural implements and the Medical Stores, which the people of this State may need, for cotton." He declared that with the consent of proper authorities on both sides, he would be glad to effect an exchange of this character to a reasonable extent, and requested for a colonel, evidently representing Brown, an interview upon the subject.[29] One day later he addressed President

Davis, explaining that he could effect an arrangement by which cotton could be exchanged for salt, if the president would allow both to pass through the lines. He does not neglect to point out that this opportunity came when salt was selling at five dollars a pound in Georgia with no prospect of a future supply; when the needs of the state were placed at over 100,000 sacks; and when he did not know whether the opportunity would be again presented. Davis, hide-bound by the law though the confederate structure was toppling about his ears, replied on the 31st that only cotton belonging to the state and to the confederate government could be so used for exchange. If certified by Governor Brown, as exclusively state cotton, permits would be issued for its exchange for salt, likewise limited to state purposes.[30]

There is even one bit of evidence which would seem to indicate that at the close of 1862 the Richmond authorities meant, at least, to tolerate trade in salt on the part of loyal citizens with the enemy in the portion of Louisiana still under confederate control.[31]

While confederate hope of success ran high during the early years of the war, thought and effort were more concentrated on trade with foreign countries to replenish the rapidly-dwindling stores of salt than was true later. With proper foresight the confederate government could have laid in large supplies of salt from abroad before the blockade was strictly enforced, for Messrs. Andrew Low and Company, merchants of Savannah, interested in or controlling several British iron clipper ships, made a proposition to the secretary of war, then L. P. Walker, on April 24, 1861, looking to that end. They offered to bring in large quantities of foodstuffs on favorable terms, but there is no evidence that the confederate authorities had equal vision; indeed, it is asserted that the proposition was not even considered.[32]

Salt did, of course, slip through the blockade. Repeatedly, as the cargo of a blockade runner is specified, we find salt enumerated.[33] The newspapers are found recording and exulting over vessels which successfully ran the blockade at one port or another. Such successes were obviously events of major importance, for we find first one paper and then another noting the fact of a successful run just as they copied reports of victorious battles. For instance, the *Wilmington Journal* late in 1861 reported that a schooner succeeded in getting safely through the blockade into the port of Wilmington with 4,500 bushels of salt on board.[34] The fact was still echoing in the press in November, when a paper at Charlotte, North Carolina, stated that the salt, which had recently reached Wilmington, sold at five dollars a bushel from the vessel.[35] Again, the Richmond *Dispatch* published the fact that the steamer *Kate* from Nassau ran the blockade into Wilmington, accompanied by a schooner with a large cargo of salt; *The Daily Mississippian* quoted the *Dispatch* a few days later.[36] A schooner from Havana ran the blockade at Mobile on August 7, 1862, with a valuable cargo of powder, lead, caps, salt, and coffee; promptly *The Clarke County Democrat* (Alabama) heralded it to its community,[37] though it is but just to add that this center of salt activity might be presumed to have had a peculiar interest in all matters relating to that article. Another vessel entering the port of Mobile in March, 1863, was duly reported as carrying in her cargo three sacks of salt, although so small an amount might have been expected to have passed without comment.[38]

The sources from which the foreign salt were imported were chiefly those from which the ante-bellum salt was derived. The steamer *Cuba,* obviously from Havana, brought into Mobile in July, 1862, about a thousand sacks of salt, over which the confederate government had full control.[39] Important supplies came from Nassau; the schooner *Convoy,* arriving at Shallosse, North Carolina, shortly before November 14, that same year, hailed from Nassau, but its cargo, 250 sacks of Liverpool salt, belonged to a firm of exporters in Augusta, Georgia, which had selected the North Carolina coast as the only point where they dared to try to run the blockade.[40] Evidently small personal ventures to run in contraband goods from Nassau were undertaken, for in November, 1862, General Whiting, in command at Wilmington, sent to Charleston three West Indian negroes, part of the crew of a small schooner, which had been beached on the North Carolina coast while trying to run salt through the blockade. As the negroes were trying to bring salt in the confederacy, General Whiting wished to encourage future activities of that sort by helping them to reach home.[41]

As was inevitable, some of the precious cargo was lost at times in the effort to run the blockade. One concrete illustration will suffice. A schooner, laden with turpentine and salt, ran into Bear Inlet on the North Carolina coast, where it was destroyed by a federal vessel, with the result that 400 sacks of the much prized salt and 500 barrels of turpentine were scattered on the beach, to await the destruction of federal sailors; on other occasions where the blockade-runner was captured, the cargo fell prize to the federal captors—but was no less lost to the confederacy.[42]

Mexico constituted another source of supply, though the salt sent over the border might often have been good Liverpool salt traveling the round-about path of Nassau, Vera Cruz, and Brownsville. In any case, whether it was domestic or imported salt, a group of Mexicans had found it profitable to bring salt over from the Pecos, while the confederate government asserted authority at the border by collecting the duties.[43]

The most serious state effort to secure a large supply of salt from abroad was that made by Governor Pettus for Mississippi. Though doomed ultimately to failure, it merits more than passing reference, because it created much comment at Richmond and for a period raised high hopes. The project was born out of the desperate situation in which Mississippi found herself for salt and apparently with no place within the confederacy from which to get a considerable supply. On October 28, 1862,[44] he presented the plan. The suggestion had been made to the governor by a general of the confederate forces that salt could be delivered by a French subject within the confederate lines, provided cotton could be sent out from the shore of Lake Pontchartrain in payment at the rate of one bale of cotton for ten sacks of salt. It was stipulated that the French consul must guarantee that the cotton would go directly to France on French vessels, not one bale to be diverted to New Orleans where it might reach United States hands. With these conditions, which he felt brought the project safely within the law, as the trade was not with the enemy or through its ports, Governor Pettus approved the plan of importing 50,000 sacks of salt.[45] President Davis gave his consent, despite the military regulations, though he thought the ratio of salt to cotton should have been higher.[46] Governor Pettus thereupon made contracts with several foreigners, one Frenchman and two British, for the exchange of cotton for salt at the rate indicated above.[47] He then directed the quarter-master general for the state, to purchase 500 bales of cotton and to have a sufficient quantity on hand to pay for the first cargo.[48] Fifty bales of this cotton was delivered to one of the contractors, M. Mennett, to be shipped to France, after he had deposited $10,000 in confederate notes with the state treasurer to secure the state against loss. But he, as well as the other contractors, failed to deliver the salt,[49] owing to change of generals and policy of the federal authorities and so presumably lost his deposit. However little the project may, therefore, have yielded the greatly needed salt, it seems to have yielded some currency, also badly needed, to the Mississippi treasury, for the last allusion in the records shows it still unreturned to the depositor.

CHAPTER XI

# SALT AS AN OBJECTIVE OF NAVAL ATTACK

WHEN THE reader thinks of the naval operations of the Civil War, he is likely to think of the exploits of the famous armed cruiser, the *Alabama,* of the other raiders which daringly and brilliantly, considering the size of the confederate navy, ranged the seas of the earth, and of the black nights, when the long, narrow blockade-runners slipped into Charleston and Wilmington past the triple lines of the federal gun-boats. It is most unlikely that he has even heard of the continuous, dogged, and very important operations of the union navy all along the Atlantic and Gulf shores from Norfolk to Vermilion Bay in Louisiana to break up the salt-making of the confederates.

It is not necessary in this connection to do much more than to remind the reader of the federal raids upon the shore of North Carolina. When the United States secured command of the entrance to Pamlico and Albemarle Sounds by seizure of the inlets, the capture of Roanoke Island was foreshadowed, an event which occurred in January, 1862, and which was followed shortly by the occupation of Newbern. Salt-making on Currituck Sound was no longer feasible. It is interesting, however, to note the tenderness with which the federal commanders regarded the prosecution of the work by individuals here as compared with the severity exercised against individuals on the Florida coast.[1] The state works were no more than created and beginning to function at Morehead City than they were so effectually destroyed by union forces as to warn North Carolina once and for all against attempting to supply her need of salt from sea-brine at that point.[2] Once more the state salt works were removed, this time to a point near Wilmington. As has been detailed in another connection, after a long period of threats against the salt works at Wilmington, the blow finally fell in April, 1864.[3] The damage, though serious, was not irreparable. It was not the extent of the damage, therefore,[4] which deterred the state from immediately repairing the works and continuing the manufacture of salt; it was the lack of adequate protection at such an indefensible point and the consequent certainty of a repetition of the destruction by federal action.

Nowhere else were federal naval attacks so frequent, so persistent, and so exclusively directed against the salt industry as on the Gulf Coast of Florida. These raids remind one remotely in their dramatic quality of the old piratical raids of colonial days in those very waters, but they were far more devastating in their results. Undisturbed for a year and a half after hostilities began, the salt works throve and waxed exceedingly great, both from the point of view of the amount of the commodity produced and of the boldness with which the industry was pursued. However, by the fall of 1862 the production had become too important to escape the watchful eye of the United States government.[5] As has been indicated in the discussion of the salt resources of the confederacy, works for the reduction of salt from ocean brine were erected along the bays and sequestered inlets of the entire Florida coast, but they found a particularly favored location on the gulf side between Choctawhatchee Bay and Tampa. Particularly, were they centered around Saint Andrew's Bay and in Taylor County; frequently they were placed at the heads of bays, from one to five miles inland from the open gulf, in order to escape the vigilant eyes of the union navy and to secure protection from the deep draught gun-boats, which could not penetrate so far up the inlets.

The first important raids on the Florida coast of which the writer has found record [6] were those by the *Kingfisher* on September 8, 1862, under the general direction of Rear-Admiral Lardner against the salt works on Saint Joseph's Bay; and by the *Sagamore* on Saint Andrew's Bay three days later. The commander of the former bark sent notice under a flag of truce to the confederate salt-makers of two hours' grace to leave the place. The confederates were seen departing within the time limit, taking off with them four cart-loads of salt.[7]

A second serious attempt followed very shortly afterwards further south, when the marines from the United States gunboat *Somerset,* reinforced by marines from the steamship, *Tahoma,* raided some salt works near Cedar Keys on Suwanee Bay on October 4. The commander of the gunboat, deeming the moment propitious for an attack, as almost all the rebel troops stationed at the terminus of the Fernandina Railroad near-by had been sent away, ran his vessel in as close as the draft of water would permit, and then threw some dozen shells until the white flag was hoisted. After the landing crew had, without any resistance, demolished a number of works, and just as they were landing to destroy the large works over which the white flag was flying, they found themselves a target for a body of some twenty-five men concealed in the rear of the building. The federal crew succeeded, despite the wounding of about half their force, in destroying several barrels of salt and a number of boats, besides capturing a launch and a large flat-boat.

Two days later, after the arrival of the *Tahoma,* a larger force landed, which, against very slight resistance, was entirely successful in its work of destruction. In the succinct phraseology of the commander of the *Somerset,* "The rebels here needed a lesson and they have had it." [8] The method pursued by the raiders in the second attack is worth recounting. The armed confederate guerillas were first routed and put to flight by the shell, shrapnel, and canister, with which the houses, woods, and underbrush were bombarded; then the small-arms men landed, deployed as skirmishers on each flank of the guns, while the working parties destroyed the boilers, which here proved to be of various shapes and curious construction. After the houses in the immediate vicinity had been set afire, the boats proceeded to the next station. At one point, it was necessary to put howitzer shells through two very thick, cast-iron kettles and through two wrought-iron boilers.[9]

November 14 a federal expedition departed from Pensacola for Saint Andrew's Bay, under orders to "run up the coast and destroy the works between this place (Pensacola) and Saint Andrew's Bay." [10] Few salt works were found until the expedition reached Philip's Inlet, not far from Saint Andrew's Bay. The extent of salt manufactories along the shores of the bays, bayous, and creeks of Saint Andrew's Bay greatly surprised Commander Hart of the *Albatross*. He noted the first day of his entry into the bay the smoke of a very large number of salt works, and when he returned on board his vessel that night, he observed that the sky was lit up both to the eastward and westward, away inland, for a long distance. He afterwards learned that the salt-makers, as they had not been molested there, had collected in great numbers on this particular sheet of water. At the town of Saint Andrews, boats which had been hauled up on the beach under sheds were set afire to prevent their "improper" use. Within a few miles of the town he destroyed several salt pans together with the furnaces, pumps, tubs, and gutters. He found some of the pans to be coast-survey harbor-buoys cut in two, each half with a capacity of 150 gallons of sea water. For about a week he pursued his work of destruction by sending armed boats in all directions. This vessel was unmolested, though constantly watched while at anchor off the town by a company of confederate cavalry of nearly a hundred men, who took good care to keep out of danger.[11]

The story of the descent at dawn on November 24 on the North Bay, an arm of Saint Andrew's Bay, with sixty marines and a working gang for destruction, may well be told in Commander Hart's own words. "The bay was very wide at this point, and a fog hung over the water, preventing us from seeing which way to go. As soon as we lay on our oars, we thought we heard voices on shore. Pulling in the direction, we soon ascertained that we were near quite a number of people, and, as we came nearer we not only heard voices, but we heard dogs barking and horses neighing, and we felt quite sure that we had stumbled upon a company of cavalry and soldiers, for day was breaking, and what we afterwards found out were canvas-covered wagons, we then mistook for tents. I thought I would startle them, and ordered a shell to be sent over their heads, and in a minute there never was heard such shouting and confusion. They seemed not to know which way to run. Some of their mules and horses they succeeded in harnessing to the wagons and some they ran off to the woods beyond as fast as they could be driven, a shell every now and then over their heads making them hurry faster. The water was so shoal that our men had to wade over 200 yards through the water, over a muddy bottom, to the shore, and before they reached it the people had all left and we could just see them through the woods at a long distance off. We threw out pickets, and Acting Master Browne, with the men belonging to the *Bohio,* took one direction, and I, with my men and officers, took the other, and, with top mauls, axes, sledge hammers, and shovels, we commenced the destruction of salt kettles and salt pans and mason work, for we had got into a settlement of salt-workers.—To render everything completely unfit for future use we had to knock down all the brickwork, to destroy the salt already made, to knock in the heads and set fire to barrels, boxes, and everything that would hold salt; to burn the sheds and houses in which it was stowed, and to disable and burn up the wagons that we found loaded with it. The kettles being such as are used in making sugar, we knew the capacity of by the marks on them, but the salt pans we could only tell by measurement, which we had no time to do, so that our total estimate of the amount of sea water that was boiling in them when we arrived is far short of which it really was." [12]

For the further data of the seven-hours' labor of destruction we are indebted to Hart's assistant. Acting-Master Browne tells of the killing of all the mules and cattle they could find, the effective method of ruining the salt by mixing it with sand, and of the fact that there were 2,500 men engaged in the industry in that immediate vicinity.[13]

So helpless did the salt workers feel themselves that at another point, four miles distant, they began to put out the furnace fires in order to save their kettles, when they saw the federal raiders appear. They abandoned, without even a struggle, over forty furnaces and sheds as well as pumps and kettles.[14] In the course of the raids, which extended to December 8, the unionists found occasion to destroy fire-wood as well as pots and pans, large furnaces which had never been used, and brick kilns with their drying-sheds. So thorough was the work that they stopped on the return trip to destroy three or four small works which they had missed earlier in the dark of dawn.[15] A working capacity at these works of 500 bushels per day was estimated.[16] As is apparent from the above recital, Saint Andrew's Bay, landlocked, with numerous arms and consequently an extensive shore-line, afforded peculiar advantages for the industry, and was regarded as a particularly sheltered and safe location. One armed vessel would have been kept busy all the time patrolling her several arms. Captain Browne estimated that the destruction had deprived the confederacy of 568 bushels of salt a day.[17]

For about six months the southern salt-makers in Florida seem to have enjoyed peace, and then on June 14, 1863, four distinct and extensive establishments located on Alligator Bay near East Pass, boasting 65 kettles, were, together with their appurtenances and huts, completely wrecked and burned to the ground by a party of engineers and marines from the steamer *Somerset.* The shelling of the woods from the vessel's guns prevented any effective resistance.[18] These works were said to be the most extensive on the coast at that time and to have been in operation since the commencement of hostilities.

A more serious expedition was that on Lake Ocala, which was combined with a new attack on Saint Andrew's Bay the next December. On the second of that month a successful attack was made on Kent's Salt Works, consisting of three separate establishments, on Lake Ocala, which lies some twenty miles west of Saint Andrew's Bay and five miles inland. The destruction of six large steamboat boilers, which had been improvised into kettles, of two large boats, six ox-carts, and the casting of a large amount of salt into the lake, meant the paralysis, temporarily at least, of works which had been producing 130 bushels of salt a day.[19]

The destruction of the salt enterprise of the confederate government on the West Arm of Saint Andrew's Bay, where 400 bushels were produced daily, followed on December 10, an act which involved a loss of half a million dollars. The government works constituted a village of some twenty-seven buildings, covering three-fourths of a square mile, and kept many hundred ox- and mule-teams constantly employed in hauling salt to Eufaula Sound, whence it was conveyed to Montgomery.[20] The burning of the thirty-two houses and shanties which made up the village of Saint Andrews was deemed necessary in order to deprive the confederacy of headquarters for the rebel military companies, which, equipped with a single fieldpiece, were protecting the numerous salt works and which were preventing contrabands and refugees from deserting to the blockading fleet. These guardsmen were effective in harassing union soldiers of the blockading fleet who landed on the main for wood.[21] The expedition then proceeded down the bay, destroying no less than 198 private establishments, which lined each side of the bay for seven miles, and aggregated a large total, although they averaged only two boilers and ten kettles each. The records reveal at one of the places raided a queer, if natural, device of the confederates of burying their kettles in order to save them. The negro contrabands pointed out to the union raiders the places in the swamp where the kettles would otherwise have been effectually concealed.[22] The damage wrought on December 11 to private and government works was estimated at well over $3,000,000; it is possible to compute the salt capacity on Saint Andrew's Bay as being reduced at this time by 1,847 bushels a day.[23]

The remarkable extent of the salt activity on this particular sheet of water is indicated by the statement from Acting-Master Browne that he contemplated destroying more than a hundred other works on the east and west arms of Saint Andrew's Bay.[24] The reason for its great popularity as a locale for the salt industry is not far to seek. The swamps in this bay were the best adapted in the entire confederacy to salt-making, as the continued drought in that region, protracted through three years, had caused the evaporation of nearly all the fresh water so that the water would test at least 75 per cent salt, whereas sea-water usually yields salt in the ratio of one bushel to about sixty gallons of brine. In addition, the salt made there was of a quality superior to that procured elsewhere, not even excepting Virginia salt. The officer directing the expedition boasted that the destruction would constitute a greater blow to the southern cause than the fall of Charleston.[25]

The work of demolition of the salt works was completed by Acting-Master W. R. Browne [26] about a week later when he proceeded up Saint Andrew's Bay against unfavorable conditions of wind, tide, and darkness, but nevertheless succeeded in destroying ninety additional salt-establishments during the hours of a long day on December 18. Furthermore, as the pickets gave the alarm, the confederates themselves set the remaining works afire to prevent their destruction by the federals—some 210 in number, for 300 salt works covered the beach for ten miles. Some other works in West Bay were also destroyed by the confederates to prevent the federals from doing so, so that the union raiding party left, congratulating themselves on having cleared at least three arms of the bay of all salt works. The total result then of ten days' work, performed by five officers and 57 men, represented the ruin of 500 works, 33 wagons, twelve flatboats, two sloops, six ox-carts, 4,000 bushels of salt, 700 buildings, and probably 1,000 kettles and iron boilers, while it would appear that the confederates destroyed an equal number of works. Subsequently the company set to guard the works disbanded, as there no longer existed any occasion for their service at that point. Adding the loss inflicted by the confederates themselves to that wrought by the unionists would bring the total by the close of 1863 to well over $6,000,000. While the officers made no computation of the salt capacity of the works destroyed at this time, they furnished statistics which make it possible of calculation. Acting-Master Browne states that there were 529 kettles averaging 150 gallons each and 105 iron boilers for boiling brine, the capacity of boilers being usually not less than 300 gallons. This figure must be doubled to include the productive capacity destroyed by the confederates themselves. The brine here was said to be 75 per cent salt, which makes 166,275 gallons of salt capacity. Translated into dry measure, 5,196 bushels would appear to have been the daily capacity of salt thus swept away from the confederacy. Not undeserving of mention is the carrying off of 48 contrabands as prisoners and the gaining of five deserters to the union service.[27]

However serious the blows struck at salt-making on the Florida coast, it is a remarkable fact that it seemed impossible to crush the industry. It was highly lucrative, but, what was more important, it was indispensable. Therefore it sank in ashes and ruin, only to rise, Phoenix-like, from its ashes. It took no longer than from December 18, 1863, to about February 7, 1864, for the smoke of furnace fires again to hang over the shores of this favored bay and for the steam again to rise from the boiling brine. It seems almost incredible that in this short space of time the Richmond government, in the throes of war and under ever-increasing stringency for money, could have reassembled the money, materials, and workers to recreate works of such stupendous size as the records show had been reestablished in about two months. Among them were some of "the largest government works ever created in Florida," according to Browne: furnaces, built of brick and stone, 25 sheet-iron boilers with an average capacity of almost 900 gallons, pumps and aqueducts, nineteen salt kettles with an aggregate capacity of 3,800 gallons, sixteen log-houses, five large steamboat boilers, one flat-boat, vats, and tanks.[28]

These salt works, covering a space of half a square mile with a capacity of probably twice 6,681 gallons of brine per day, or 313 bushels, estimated in value at not less than some hundreds of thousands of dollars,[29] had been in operation only ten days when detected by the ever vigilant Browne. He sent off a force in two detachments on the morning of February 17, one to march inland seven miles in order to attack from the rear, while the other on the cutter should attack from the front simultaneously. Success again crowned his efforts so that the destruction must have represented a loss which the confederacy could ill afford to sustain, amounting to hundreds of thousands of dollars.[30]

During the same month [31] two expeditions were sent from the steamship *Tahoma*. During the three-day period, February 17 to 19, the first expedition destroyed the government works near Saint Mark's Bay, which were so extensive that they required a day and a night of unremitting toil by two detachments of men before the work of demolition was accomplished. A long hard march of forty-five miles through swamps and dense woods marked a new feature in methods. On the twenty-seventh, a week later, a second expedition, cleverly eluding a rebel cavalry company stationed to protect the works, destroyed government works at Goose Creek. The confederate government thus lost by these two raids a capacity of 2,500 bushels per day.[32] The list of property destroyed in the first expedition on Saint Marks is overwhelming and was quoted as worth $2,000,000, while the later destruction, covering such items as 98 well-constructed brick furnaces, 165 pans, 100 store-houses and 2,000 bushels of salt, was obviously very great. Perhaps an estimate of $3,000,000 in all is no exaggeration.[33]

April 12, 1864, saw the ever-active Browne, as befitted the commander of the bark *Restless,* operating in the East Arm of Saint Andrew's Bay. He again fell upon government salt works, which seem truly, as Browne complained, to be rebuilt as "fast as he demolished them." [34] Browne's zeal outran his judgment when a few days later he wrote to Admiral Bailey on April 22, after destroying some large works, "There are no other works on East Bay at present, and I hardly think they will attempt to make salt on this bay again, as they are aware of our having mounted a howitzer on the barge. There are some few small concerns in operation on West Bay, and I am only waiting an addition to those now there when I shall fit out another expedition and am in hopes of making quite a formidable raid, one that will make it worth our while to report." [35] Certainly it seemed hopeless to extirpate the industry when after such pitiless, unremitting destruction an ensign was obliged to report to Browne on May 26, 1864, his belief that there were still, judging by the fires, 200 salt works on Saint Andrew's Bay. He had learned from good authority that the salt-makers had a large quantity of new boilers on hand to replace the broken ones as soon as possible after the federals left the bay.[36] But nothing could dampen Browne's ardor against salt works, for on May 27 upon receiving his ensign's report he registered a determination to destroy that work, and, "if possible to do so as fast as they build new ones." [37] This vow he lived up to on June 8, when ninety-seven works with their equipments were totally destroyed, and on October 5 and 6, when kettles and boilers on that same bay with an aggregate capacity of about 50,000 gallons were demolished.[38]

The concluding scene in the long-drawn-out drama which had Saint Andrew's Bay for its setting was not played until February, 1865, when the last demolition of salt works on West Bay was staged and pans with a boiling capacity of 13,615 gallons broken.[39] Meanwhile extensive works on Rocky Point in Tampa Bay had been destroyed by the federal fleet during the preceding July (1864), while in December some works had been swept away by detachments from the United States steamers, *Stars and Stripes, Hendrik Hudson,* and the *Ariel.*[40] Admiral Bailey also sent an expedition in September, 1864, to swoop upon the very large salt works at Bon Secour Bay, which embraced fifteen houses, each producing 75 bushels a day, which made a total daily production of over 1,100 bushels.[41]

Such measures to punish any effort of the confederates to relieve their salt famine led, naturally, to counter measures to beat off the attacks from the blockading fleet and to protect the salt-makers. It is easy to trace the first repercussion from the raids on Saint Joseph's and Saint Andrew's Bays of early September, 1862, because toward the close of the month we find Governor Shorter seeking the help of the central authorities. Twice, he complained, had Alabama citizens, who had for several months been making salt for their personal use on the Florida coast from Saint Joseph's Bay to Choctawhatchee Bay, been interrupted by federal raids. He estimated that three hundred armed men would afford the requisite protection.[42]

Promptly on the heels of the serious raids of December, 1862, the Florida legislature made provision for the organization of the salt workers, citizens and non-citizens of Florida, for defense against the raiders under officers appointive by the governor. All salt workers must be enrolled for military duty, and were subject to call in the face of an apprehended invasion of the enemy; refusal to enter such organizations would be met with prohibition of the right to make salt. Such service did not, however, carry any compensation.[43] The law was promptly translated into action so that by January 9 of the new year the officer appointed to the work was able to report the organization into companies of all the salt-makers on the coast between Saint Mark's and the Suwanee River, but among the 498 men physically capable of duty, there were but 43 guns in good condition and very little ammunition. Places of rendezvous in case of an attack had been indicated, while the salt workers had been ordered to cooperate with the military in their vicinity.[44] Meanwhile General Cobb, in command of that military district, took prompt steps to strengthen the position. He had found the few troops stationed there too far from the coast to render timely aid; he therefore shifted them to more convenient posts and took steps to concentrate the operations of salt-making within narrower territorial limits.[45]

Governor Shorter of Alabama thought to help the situation by authorizing Colonel Clanton, a dashing, experienced, and indefatigable Alabama officer, to raise a regiment of cavalry rangers for six months.[46] That officer was hopeful that if he mustered in two or three companies of infantry and could get a section of artillery, he could drive the enemy to their gunboats and give security to the salt manufacturing industry.[47] On the whole it was true, as a citizen of Tallahassee said, that the arrangements for defense of the great salt interests were merely such as were incidental to the general protection and observation of the coast. Cavalry companies were stationed at Newport and Blue Creek in Taylor County within two or three miles of the salt works in that county. Salt workers on the coast of Georgia were likewise dependent on confederate cavalry in that section and on the troops for border defense.[48] The proposition of concentrating the salt workers within narrow territorial limits could hardly be adopted without damage to the enterprise, as it would promptly increase the difficulty of obtaining wood and supplies. A further obstacle to such concentration was presented by the fact that much of the coast land was owned by individuals, who would in the disposal of locations consult their own interests.[49]

Finally, the governors of Alabama, Georgia, and Florida united in requesting of the Richmond authorities a new military department to be composed of portions of their respective states, offering as one argument for the move, their extensive seaboard, along which thousands of their citizens were engaged in making salt for themselves and for their neighbors hundreds of miles in the interior.[50]

The blockading fleet was, naturally, constantly alert to detect fresh evidences of confederate activity in salt-making on the coast, as that enterprise afforded a particularly vulnerable point of attack. On every coast there was need of vigilance. On September 26, 1862, very extensive salt works, were detected near Bluffton, South Carolina, up the Bluffton River, an inlet running far inland near Port Royal Sound and hence well protected against any except very light draft boats.[51] They were completely demolished with all their kettles, furnaces, and appurtenances. Though no attempt was made to estimate their yield, it was stated to be very great.[52]

The mishaps which could easily prevent the success of the federal raiders are so well illustrated here that the details should not be omitted. Starting from the post of the blockading fleet, Fort Pulaski, Georgia, at one o'clock at night with two armed steamers, which carried five companies on board, the commanding officer arrived at the mouth of the river three-quarters of an hour before daybreak and proceeded up the river as rapidly as possible. His design was to land the infantry force one and a half miles below Bluffton in order by a rapid march to gain a point to the rear of the village where the roads converged, thus cutting off the squadron of cavalry stationed there. The *Planter*, unfortunately, ran aground in the thick fog when within a half-mile of the point aimed at so that the noise of the engines in getting her off betrayed her to the confederate pickets. The troops were, nevertheless, landed as soon as possible, while the steamers proceeded directly to Bluffton, keeping their distance about half a mile ahead of the troops on shore. The village was remarkably protected against an attack by water, since bluffs entirely commanded the narrow channel through which vessels necessarily approached. There was every evidence at the village of the precipitate departure of the cavalry and of a portion of the inhabitants. As soon as the infantry had come up, the commander of the expedition pushed on up the river to several plantations where he had learned there were extensive salt works. Immediately after leaving the wharf he detected some 200 cavalry and a body of infantry retreating, upon whom he opened fire from all his guns until there was soon a perfect stampede among them. After the furnaces and vats at the first point had been demolished, he proceeded to other salt works just above, which were pointed out by some contrabands who had, as usual upon the arrival of union forces, made their appearance. The yield of salt at the last location must have been very great, for the vats extended for more than a quarter of a mile. Meanwhile the confederates again made their appearance on a high bluff a mile distant with a piece of artillery, but beat a retreat upon the discharge of the federal guns.

About a month later, around November 3-10, a small force of negro refugees traversed many miles of the rivers and bays of East Florida and Georgia, between Saint Simon's Island and Fernandina, destroying many salt works, together with a large number of teams and wagons. This coast was, in a sense, a salt-maker's paradise, for the entire shore is intersected by bays, lagoons, and rivers, penetrating for more than one hundred miles inland into the heart of the country and navigable only by light-draught steamers. Marshes and thick woods line the banks of the streams. A distinctive feature of this expedition was the fact that it was conducted with a company of colored troops to test their fighting qualities. The pickets were driven in and the work of destruction completed before the enemy could collect a sufficient force to overpower the raiding force. "Rarely, in the progress of this war," according to the commanding general, "has so much mischief been done by so small a force in so short a space of time." [53] In all, it cost the confederacy nine large salt works, together with $20,000 worth of horses, salt, corn, and rice, which the raiders could not carry away.[54] Later raids on salt works, incidental to other expeditions, carried union men into the interior of Georgia by hard marches across the swamps several times during the year 1863. In February they pushed up Crooked River, and in early June moved against some works seven miles above Brunswick.[55]

In Princess Anne County, the southernmost shore of Virginia, northern zeal in January, 1863, discovered and destroyed some salt works at Land Bridge, a point fifteen miles below Cape Henry, and an even larger number fifteen miles further south. At the two points together over 1,400 bushels of salt were rendered useless in the usual ways. The expedition then proceeded to Wales Neck, where it destroyed a large lot of pans and lumber, evidently designed for the construction of salt works, as well as 500 cords of firewood.[56]

At the other extremity of the confederate coast, the same work of destruction was going on. One reads of an attack on an extensive and valuable salt manufactory at a place called Cedar Lake in Texas on November 27 and 28, 1862. The salt ready for the market, amounting to about ten tons, was here found packed for transport in hide-bags instead of in the weak osnaburgs familiar elsewhere. When the unionists engaged in their labor of destruction at another point, a few miles further down the coast, a mounted troop of guerillas charged on them at the same time that a large body on foot crept up through the grass to drive them to their boats, with the result that only one factory was even partially destroyed.[57] Thus was waged all along the coast the contest for the right to the use of the brine of the sea, and in the contest the unionist forces were in the end pretty generally successful in their denial of that right.

## CHAPTER XII

# SALT AS AN OBJECTIVE OF MILITARY ATTACK

UNTIL ONE embarks upon the study of a relatively insignificant subject, one has no conception of just how such obscure or unimportant factors can determine the military strategy and battles of a great war. The sources of the salt supply for the confederacy—the Goose Creek Valley, the Kanawha Valley, the mine at New Iberia—were repeatedly the objectives of attack, while constantly, and again above all, the Holston Valley of Virginia was never out of the picture.

The salt works at Fishing Creek, West Virginia, were the first attacked and captured. A raid in January, 1862, resulted in nothing more serious than the breaking of kettles and pumps.[1] It remained for another union officer to do a much more thorough piece of work in dismantling the five works on Goose Creek the next October. A brigade was ordered forward to Manchester on October 21, 1862, and did its work of destruction on the twenty-third and twenty-fourth. Over five hundred men [2] worked in details for thirty-six hours in the effort to render the works ineffective for future use. The report of the officers in charge recites how they destroyed the pumps and wells and even the pipes conveying the brine. In every case the pumps, or such portions of them as were removable, were taken out of the wells, broken into pieces, and the fragments forced back into the wells; in two wells, a piece of iron grate was rammed into that portion of the pump which could not be drawn out; in another instance, an iron bar was forced into the well-pipe. Three hundred feet of surface pipe, the scaffolding, and all the wood-work of the pans were destroyed. At two of the works, after forcing back into the wells such portions of the pumps as were removable, cannon balls were driven in. At all the works large quantities of manufactured salt, estimated in a hasty examination at about 30,000 bushels in all, were found and destroyed. The buildings, because they were regarded as valueless, were not disturbed. With the permission of the general in command, loyal citizens of the neighborhood were allowed to remove enough salt to supply their immediate needs, under a binding oath not to permit any of it to be used for the benefit of the confederacy. This privilege was regarded as necessary to prevent great suffering among the civilians; furthermore, the salt works were located in the midst of a population loyal to the union.[3] As the confederates had already carried off possibly 4,000 bushels, and every circumstance led the raiding party to believe that the large quantity on hand would have been hauled off as soon as transportation was available, the remainder of the salt was destroyed by turning the water from the cisterns on it, or by throwing it into the pools and streams nearby. The destruction was justified on the ground that an army sufficient to protect the works could not be subsisted from the surrounding country, as bad mountain roads and the severity of winter would have precluded the hauling of supplies. The commanding officer reported that the works would be ineffective except after a considerable period of time and a large expenditure of money for the boring of new wells.

The destruction, however, was possibly less thorough than the officer fondly hoped, for only a few days later we find a colonel writing President Davis that the Goose Creek salt works must be put into operation since coal was within convenient distance and especially since it would secure the control of the important article, salt, in the hands of government agents—a result which he declared worth more than 5,000 troops.[4]

The confederacy sustained a double blow at its salt resources at just about the same time, for by November 1 of the same year, General Echols had been obliged to abandon the Kanawha Valley and West Virginia. Indubitably, possession of the salt works in the Kanawha Valley was of the first importance to the south. Certain persons held that the salt from this source alone was sufficient to supply the whole confederacy. Since these works had been largely relied on for the Cincinnati trade, the south expected the union to strike for them to insure its own supply of salt.[5] At an early period of the war strong efforts were made by the confederates to enter and hold western Virginia, as the great value of that fertile country as a source of food supplies and horses for the army, as well as its importance for coal and salt, could not be overlooked. The federal troops, shortly after the western counties of Virginia had organized into the separate state of West Virginia, entered from Ohio, as is well known, and under General Cox marched through the Kanawha Valley in July, 1861. The close of the very first year of the war saw the confederate army driven, temporarily, at least, from the state.[6]

It would appear, however, that the union army did not maintain a sufficiently large force to hold West Virginia, for we find the confederates in the fall of 1862 first rejoicing over the recovery of the Kanawha Valley [7] with the valuable salt works by General Loring, and then bewailing their loss. General Loring, acting under orders directly from Secretary Randolph,[8] began his campaign by moving rapidly from Giles Court House [9] in southwestern Virginia on September 6, 1862, with a force of 5,000 men upon a raid which ended ten days later. He succeeded, despite the efforts of the union forces to burn all the furnaces, in capturing the salt wells virtually uninjured,[10] together with a large store of salt;[11] the negroes, however, by whom the work had formerly been conducted, were carried off by the enemy.[12] Loring drove the unionists clear back to the Ohio River.[13] Virginians were keenly interested, for as early as June, several months before the project was underway, Governor Letcher issued a proclamation, appealing to citizens of his state to volunteer under General Floyd to recover the daughter state. That the salt works were a prominent factor in the project is evidenced by the fact that in addressing the legislature on September 20 he stated that General Floyd was on his way to Charleston, and added that the object in raising the state force was to protect the Kanawha salt works by a suitable force.[14]

By October, the unionist leaders in their turn were interested in confederate activity at the salt furnaces, so that we find Governor Pierpont [15] warning Secretary Stanton that the confederates had a force of 10,000 troops and forty cannon at Charleston, working and transporting salt south.[16] This was an open recognition by the latter of the fact that they hardly hoped to retain the section.[17] Before October 8 President Davis had warning that his foes were fitting out an expedition to move down from Pittsburgh for the recovery of the salt works.[18] By October 15 a change in command had placed General Echols in charge of the defense of the Valley, which he was urged to make his first object.[19] November 1 Echols began to fall back from the Kanawha before General Cox, who was advancing with a federal army of 8,000 men.[20] That gossipy war clerk in the War Department to whom we are indebted for so many valuable hints recorded sadly on November 9 the abandonment of Charleston and its neighboring salt works because of the inadequacy of the force to hold them. "The President," he philosophizes, "may seem to be a good nation-maker in the eyes of distant statesmen, but he does not seem to be a good salt-maker for this nation. The works he has just relinquished to the enemy manufactures 7,000 bushels of salt per day —two million and a half a year—an ample supply for the entire population of the Confederacy, and an object adequate to the maintenance of an army of 50,000 in that valley. Besides, the troops necessary for its occupation will soon be in winter quarters, and quite as expensive to the government as if in the valley. A Caesar, a Napoleon, a Pitt, and a Washington, all great nation-makers, would have deemed this work worthy their attention." [21]

Although a few raids were made subsequently into West Virginia by the confederates, the most serious of which was that of Generals Imboden and Jackson in April and May, 1863,[22] no movement seriously threatened again the federal hold upon these salt works in West Virginia.

The joy of the confederates over the gift of a salt-mine of illimitable size straight from the gods was short-lived, for in less than a year it had been wrested from their grasp. It will be recalled that it became known to Judge Avery on May 4, 1862, that he was the owner of a mine of solid rock salt, and that it was felt necessary in the interests of the entire confederacy for General Taylor to take general supervision. By November we find that officer so anxious concerning the threats of the federals that he was appealing to his superior, General Pemberton, for a regiment of "reliable" infantry on the ground that the salt works might be endangered.[23] Yet somewhat inconsistently he was at pains to explain a few days later that he had no reason for the moment to apprehend interferences with the use of the mine. He pointed out that water communication between the mine and the Cis-Mississippi region by means of a short portage from the Teche to the Atchafalaya should be preserved, but in the same breath he emphasized the difficulties of defense in the Louisiana topography by virtue of its many navigable streams.[24]

General Butler's first effort to take the mine met with little success. In January, 1863, he sent a steamer and two gunboats to Vermilion Bay, one of which anchored within two miles of the salt works, though the *Saint Mary* could not approach within five miles. That night the wind veered around to the north, drove the water out of the bayou, with the result that the two gun-boats were left on a mud-bank, from which they were removed in the course of fifteen or twenty days.[25]

When General Banks was ready to move for his operations in West Louisiana, the capture of the mine came easily without let or hindrance from the rightful owners, as the inadequacy of any defensive force necessitated its evacuation, though removal of the guns was effected. On the morning of April 17, 1863,[26] Colonel Kimball advanced upon the works south of New Iberia, destroyed the eighteen buildings, with their steam engines, windlasses, boiling and mining equipment, together with 600 barrels of salt which stood ready for shipment. With the blowing up of the magazine and the removal of one ton of powder, his task seemed easily achieved.[27] The inability of the confederates to put up a fight for one of their most precious possessions, a strategic and vital spot, is tragic evidence of the weakness of the cause by this date.

If the salt works on the Florida coast and the mine in Louisiana were worthy of such pointed military attention as they received, southwestern Virginia was a region of the highest strategic value. Its possession was a matter of the first consequence to the confederacy, not alone for its salt deposits, but also for its lead; but its seizure and control was of almost equal concern to the union in order to weaken the opposition by the loss of these vital resources for prosecuting the war.

The geography of the section determined the method of the warfare for its possession. As the mountainous nature of the country rendered operations by a large army impracticable, numerous invasions by small forces, principally of cavalry, were made to destroy the salt works and the railroad. The broken surface made it difficult of defense, and rendered necessary a larger force of occupation than was usually available. The main approaches to the works from the west were through two gaps in the mountain ridge of northeastern Kentucky, Pound and Louise gaps, through which dashes were possible from Kentucky and Tennessee. A further means of crippling the foe was afforded by the single track of the Eastern Tennessee and Virginia Railroad, over which tenuous line must enter provisions for the salt-makers and over which the salt must be shipped out. After the loss of the Goose Creek and Kanawha works, the southerners were peculiarly sensitive to any threat directed at southwestern Virginia, as they fully realized that such movements jeopardized their only remaining large source of salt supply.

A list of the threats made by the federal officers against this section is a lengthy one, and yet it is a remarkable fact that it was never raided until toward the very close of the war. Solicitude concerning safety of the Holston Valley seems never to have been absent from the minds of southern leaders. "Will he," speculated Brigadier-General Marshall to Adjutant General Cooper on November 12, 1861, referring to the enemy, "pursue Williams (a confederate officer) to Pound Gap, or will he let Williams go, and come through what he supposes to be an unobstructed country down upon this road, pass Buchanan County, descend upon Richland, (20 miles from this point), destroy the salt works, and the railroad between Wytheville and Abingdon, and hold a position at Richland until reenforced? This can be done, sir, by 300 cavalry, who would be perfectly unobstructed, and who could ravage all the country to the vicinity of Wytheville. The loss to us would be irreparable, for the whole Confederacy depends upon these salt works greatly." [28] He concluded that it would be proper to avail himself of the state militia, for which certain Virginia counties could be depended upon to the number of 1,300, at least until the enemy was pressed back beyond the gorges of the Mountains. He repeated his request for another battery, and he pressed urgently for another regiment or two "from other important fields of service" until he should have time to call for two regiments of the new Tennessee levies.[29] He conscientiously kept the salt works covered until the danger had passed.

This same officer displayed the same lively concern a few months later. In February, 1862, he had, he declared, positive information that the enemy was contemplating a movement into Virginia along the line of the mountains, that he was already using the navigation of the Sandy River to collect supplies at Piketon,[30] and that he was collecting his troops there as rapidly as possible, with the intention of bringing them up to the number of 12,000. From the rumor that the Pound Gap and the Wytheville road were to be the paths of invasion, he argued that the salt works and the railroad were the objectives.[31] Again concern seems to have been based on a rumor and nothing more.

General Loring also proved nervous when the responsibility for the defense of the mines, salt manufactories, and railroad devolved upon him in May of the same year. He complained to both Cooper at Richmond and to General Lee that the force to guard the interests in that section were entirely inadequate and fixed five regiments and a battery as the minimum requirements.[32] Needless to say, the confederacy had no such regiments to spare from points under attack for a remote mountain section, however strong was the argument that it might be attacked; even though one of its generals was falling back from Lewisburg in West Virginia, many mountain ranges interposed between him and a pursuing foe.

During 1863, we find the command, and hence the anxiety for the section, transferred to General Sam Jones who was in command of the Department of West Virginia, and, more immediately, to General Floyd, who was stationed at Dublin, not far from the salt works, in command of the Virginia State Line troops. We find General Jones not only alert to warn Floyd of projected movements,[33] but also offering constructive suggestions for strengthening the defenses. "It would be well," he wrote Floyd on January 29, 1863, "to have a few defensive works. Block-houses for artillery and infantry would be best, on commanding points near the salt works. The employees of the works could, if organized and armed, use them with good effect in defending the works. But I have no force at present to construct the works. If you will furnish the labor, I will send an engineer officer to locate and superintend the construction of the works."[34] Administratively, the defense was not well organized at that time, as the salt works lay directly on the boundary line between two military departments, so that the Virginia counties through which lay the most practicable approaches were not under the care of the officer responsible for the works.[35] Furthermore, Jones promised to send 1,200 to 1,500 men to Saltville in less than twenty-four hours, if he were given early notice of the advance of the enemy, while artillery at a point not far distant, Jeffersonville, could be brought up in time to help in the defense. Differing also from other military men, notably Whiting at Wilmington, he had a due regard for the importance of the salt production, so that he was unwilling to impress the salt hands except under the most convincing emergency.[36]

A slight flurry of excitement agitated these commanders about April 18 to 28, when the information spread that the enemy, 2,500 strong, on Clinch River near Moccasin Gap was believed to be moving on the salt works. Jones instantly hurried 2,200 troops to their defense with two field batteries, with which it was felt that the mountain passes could be held until reenforcements arrived. The commanding officer at Saltville was directed to hold the passes leading to the salt works "at all hazards." [37]

One of the tales of feminine daring and patriotism, which are not at all uncommon in connection with this war, has come down concerning Saltville and Wytheville. In late July, 1863, Colonel Toland, aiming at the sources of salt and lead, invaded the region and had reached Peery Farm near Jeffersonville (now Tazewell). The news of his approach spread swiftly till it reached a young eighteen-year-old girl, Molly Tynes, who slipped out through a rear door so that her plan should not be thwarted by an over-cautious family, saddled her own fleet bay, and raced off for Wytheville. Her journey meant the crossing of five high mountain ranges, the traversing by night of a wild region, infested by wild animals, and the making of her own path, when the bridle path ceased, through tangled undergrowth. She spread the alarm as she dashed by, and reached Wytheville at dawn so that a defense guard, made up largely of old men and boys, was hastily organized to face and rout Colonel Toland when he rode into the outskirts of the town a few hours later. Toland fell, and his second in command, Major Powell, was captured.[38]

While the threatened raids against the salt works usually did not materialize, the experiences seem to have convinced General Jones that a force of from 800 to 1,000 men, under an intelligent and determined officer, should be assigned to the permanent defense of the salt-works, together with eight to ten field pieces of artillery, and at least one company of cavalry. Furthermore, an engineer officer should be sent to plan and superintend the construction of a few defensive works. Such a force, in his judgment, should be able to hold back three times their own number of the enemy and would allow the commanders of the departments greater freedom of action.[39]

Either before or at about this time General Jones sent his chief engineer to Saltville to plan and lay out such defensive works as he thought requisite, while the general also secured from the chief of ordnance six field pieces to be placed in position there. The War Department, however, sent another chief engineer there from General Buckner's staff for the same purpose, with the result that two plans, differing materially, were devised. Thereupon Jones recalled his officer, preferring to leave the matter to the other military department.[40] Meanwhile, he had taken thought for the immense quantity of salt, much of it government property, which was accumulating at Saltville daily, and on July 17, 1863, urged on the commissary-general that it be distributed along the line of the railroad as a precautionary measure against the raid which he anticipated would soon occur.[41] Cavalry also for rapid movements was, naturally, an essential factor in the defense. In September General Jones is found pleading with Governor Letcher for authority to impress slaves from the adjacent counties in order to complete the defensive works. When Letcher failed him, he turned to Secretary of War Seddon,[42] but this question was at once engulfed in the more important one of troops to withstand the serious attack, which menaced the region from September 9 through most of October.

The enemy was reported moving from Cumberland Gap toward Saltville, and so again all was excitement.[43] General Jones telegraphed Seddon that he could not answer for the safety of the place unless reenforcements were sent. This time the attack was no idle threat, for the federals advanced seven miles beyond Bristol to within 35 miles of Saltville, pressing General Jones vigorously with superior force and threatening one of his officers from a strategic position between the two confederate detachments. The Sixtieth Virginia Infantry and Eighth Cavalry were ordered to Saltville; the small force at Wytheville was summoned to Glade Spring, and the home guards of Montgomery, Roanoke, Botetourt, Bedford, Wythe, Smyth, and Pulaski Counties were called on to support the defense at Glade Spring, but could not muster more than 700 to 800 men. However, after a skirmish at Zollicoffer on September 20, the federal forces moved off in the direction of Blountsville and Jonesborough, thus relieving Jones for the moment. The danger does not seem to have been over, however, until toward the end of October, for we find an urgent appeal on October 15 to neighboring villages for "every man who can possibly bear arms to go to Abingdon immediately," as the emergency was great.[44] The danger seems to have passed without attack on the works.

The gestures, on the one side, and the nervousness, on the other, continued all through the year 1864. As early as January General Grant requested General Foster to send forces to Abingdon with a view of destroying the road between Abingdon and Saltville in order to prevent the shipment of salt.[45] The accomplishment of that work, would, in the former's opinion, be worth considerable risk.[46] Still in February he had the same thought in mind, while issuing directions to another officer. To Schofield he wrote on February 16, "The destruction of important bridges between Bristol and Saltville and of salt works there would compensate for great risks." [47] That general replied encouragingly a few days later that if General Meade would, at the proper time, occupy the attention of Lee, he, (Schofield) might be able to reach the salt-works. "I propose to go prepared to reach the place if possible. At best it will exhaust all my resources, and it will be impossible without "a diversion in Virginia," as he, the writer, knew that it would be well defended.[48]

Federal officers seemed to be watching closely the strength of the defense at Saltville, which fluctuated, of course, with the danger of an attack, but which was seldom a large body, considering the interests involved. In May, 1862, General Frémont reported the number at Abingdon and at the salt works as estimated at 7,000;[49] in February, 1864, only 4,000 men were reported at Saltville, but the works were felt to be strongly fortified;[50] in April, 1864, scouts were reporting to General Schofield only about 3,500 men, mostly cavalry, left to protect the salt works.[51]

Allusions to the salt works are seldom long absent from the communications of confederate officers in that area. In March, 1864, we find General Breckenridge charged with their protection and soliciting the cooperation of Longstreet in guarding the mountain approaches, but insisting on a distinct understanding as to whose responsibility they were, since the poor drawing of the lines of the military commands still seems to have left them on the dividing line between his and Longstreet's territory.[52] If an emergency drew a large part of the forces from Saltville to meet a more serious threat elsewhere, the men were promptly returned after the danger at that point had passed.[53] On the other hand, the union officers were clever enough to know that feints in the direction of Saltville tied up confederate troops there and prevented relief from being sent to other points.[54]

Two somewhat serious efforts were made by the federals in early May and later in October, 1864, to take the salt works. On May 8 the confederates learned of the simultaneous advance of two strong federal columns, one under General Averell, threatening the location with a body of cavalry, and one under General Crook, who was threatening the communication with Richmond by destruction of the all-important railroad. General Morgan hastened 400 dismounted men of his command to support Jenkins at Dublin Depot. Meanwhile Morgan, by virtue of astutely reading his adversary's mind, was able to force Averell back from the attack on the Wytheville lead mines.[55] Late in September General Stephen Burbridge advanced through Eastern Kentucky with 5,000 men against King's Salt Works. Colonel Giltner advanced with a small brigade of cavalry of not more than a thousand men to oppose Averell; by dint of clever strategy and stubborn fighting the former succeeded in detaining Burbridge two days, until, when he was finally able to arrive at the salt works, a force equal to his own under Breckenridge stood ready to oppose the federal army. On October 2 Burbridge attacked, but the boys and old men fought desperately before they were driven back. Fortune fluctuated from side to side during the entire day, but night found Burbridge out of sight of the salt works, while the next morning he was twenty miles away and obliged to withdraw.[56]

The federal forces were, however, at last serious and the real blow fell in December. The expedition, under command of General Stoneman, consisted of a body of about 4,000 men, hastily called together in Kentucky and mounted in a surprisingly short space of time, and of a portion of a Tennessee cavalry brigade. It left Knoxville, Tennessee, on December 10. No one, not even the officers, knew the objective, and it was the third day after their departure before the enemy even learned of the movement. The department had been drained of confederate troops so that Breckenridge had only about 1,000 to 1,500 troops, which handful he concentrated at the salt works. Stoneman occupied Bristol and Abingdon, after some small skirmishes, tore up the railroad, and then, changing his tactics, passed by the salt works to advance on Wytheville where he completely destroyed the lead mines. He then wheeled on Breckenridge, who had moved out from his base to a small village named Marion in the hope of arresting the federal advance.[57] The latter, cut off from the salt works, fought an engagement on Sunday, December 18, which resulted in a partial victory for the confederates. But during the day Stoneman had sent a detachment down another road to the objective, now left defenseless except for a few teamsters and 700 home guards under Colonel Preston, who were easily captured or driven into the mountains. If Breckenridge had remained within the fortifications surrounding the works, which were well defended by eight cannon, it would have been difficult, if not impossible, Stoneman admitted, to have taken the place. The former committed the mistake, for which Stoneman had hoped, of sallying out to follow the federals. When Breckenridge found himself at night nearly surrounded and cut off from Saltville, he escaped over the mountains into North Carolina.[58]

Ultimately the long desired salt wells were in the hands of the federal forces. The old tale of destruction, so familiar by this time to the reader, was repeated: furnaces, kettles, engines, pumps, and machinery were broken into pieces, the wells and shafts filled up with shells, spikes, nails and old railroad iron, the buildings burned, and a store of salt, variously estimated at from 50,000 to 100,000 bushels ruined. The whole of the day of the twenty-first and of the night following large wrecking parties armed with sledge-hammers were kept employed at the work of demolition until, in the words of Stoneman, "a more desolate looking sight can hardly be conceived than was presented to our eyes, on the morning of the twenty-second of December, by the saltworks in ruins." [59] However, Burbridge's proud boast to Secretary Stanton that the expedition would be "more felt by the enemy than the loss of Richmond" seems scarcely to have been justified,[60] for an inspection of the works a few days later brought the report to Breckenridge that not two-thirds of the sheds and not one-third of the kettles had been destroyed, while some of the sheds and furnaces were left untouched. He hence concluded that the loss was not nearly so serious as at first apprehended. Mr. Stuart of the famous firm of Stuart, Buchanan and Company, thought no injury which a raiding force could inflict would suspend operations longer than for a few weeks.[61]

Meanwhile, no picture of the waste of war involved in the single item of salt is complete if we stop here, however sickening has been this recital of the wielding of sledge hammer and the mixing of valuable stores of the scarce commodity with sand or mud along the coasts and at the great centers of production. Throughout the *Official War Records* we come upon the statements of destruction of every little store of salt which might be of benefit to the confederate cause, unless, indeed, the federals had use for the article themselves. August 23, 1862, a union officer reported duly as a result of a skirmish with the enemy the capture of eight horses, twelve swords, fifteen rifles, and eight barrels of salt, which had been smuggled through the mountains; [62] in November, 1862, a colonel, reporting on a recent raid, records success, because he destroyed, along with a distillery, two dwelling-houses and outbuildings, several barrels of spirituous liquors and about fifty barrels of salt.[63] A New York officer, participating in the general expedition through western Louisiana, which captured the salt-mine in April,

1863, left a thousand barrels of salt uninjured because they were close to the Bayou and could be easily sent down the Teche River.[64] One federal officer, who had captured on the Red River, three confederate vessels laden with supplies for the southern army, including 200 barrels of molasses, 10 hogsheads of sugar, 30,000 pounds of flour, 110,000 pounds of pork, "and a large quantity of salt," burned the ships when he found that his booty endangered his prompt return to his base. "We had not time to transfer their cargoes," he states briefly.[65] Hurlbut, commanding at Baton Rouge, reported as one of the features of the capture of Camp Moore on October 5, 1864, the destruction of 2,000 pounds of salt, along with 4,000 pounds of bacon, 100 dozen pairs of boots and shoes, and 2,000 sides of leather, besides other stores.[66]

When the tale was reversed and the captures of stores were made by confederates, we are not so likely to read the sequel of destruction, though that procedure was followed, of course, rather than let the foe derive any benefit. More often, however, there can be read between the lines joy in the seizure of "an abundant supply of salt" as the tangible fruits of an expedition.[67] And so the sickening tale of destruction went on ad infinitum, enlivened by occasional touches of humanity as when General Howard, who was in command at Columbia, South Carolina, in February, 1865, ordered the chief commissary, before the store buildings were destroyed, to furnish to the Columbia Hospital as much salt as was necessary, and to save the surplus for the poorer citizens who had been burned out.[68]

## CHAPTER XIII

# SALT AS AN ITEM IN THE BUDGET

In any study of this subject which pretends to be reasonably comprehensive the questions of the degree of success achieved by the confederates in attaining their goal of an adequate supply of salt as well as the price which they paid for that supply are questions of the first importance. No scholar would presume to present authoritative figures. Obviously the reports cannot be complete, as there was much salt produced by individuals for their private consumption and by small makers for sale which never found its way into any records. Likewise, there can be found no records of the cost at which this private salt was produced or of the extortionate prices often paid by those who had the wherewithal to pay, though obviously the cost of manufacture per pound was much higher than for that portion of the commodity which was made in large amounts by the states or by the few large firms, such as were then operating in Virginia. The further fact must be pointed out that even for the salt produced by state enterprise and for governmental use, the records are sadly incomplete. For instance, the writer has been able to find only a few of the contracts into which the confederate government entered, notably with Stuart, Buchanan and Company but also later with the state of Virginia and with lesser firms. Likewise, the records are entirely missing for some of the states; almost nothing seems to have been preserved on the subject in Tennessee or Texas, very little in South Carolina and Arkansas, and what little survived the post-war years in Florida seems to have been lost recently. On the other hand, one is cheered by the specific, if physically appalling,[1] financial records which are to be found in the archives of Alabama and Mississippi, and also in condensed form in Virginia [2] and North Carolina.[3]

However greatly statements on the statistical and financial side must be qualified, however earnestly the reader must be warned against easy generalization and false conclusions from partial data, it is conceived that such presentation of the subject as the available material permits is slightly illuminating on this important phase of the subject, and therefore better than total silence. In other words, the writer will proceed on the tacit assumption of the reader's acquiescence in the view that "half-a-loaf is better than no loaf at all." The data afford at least some conception of the degree of success and of the cost of that success in monetary terms.

First of all, the salt needs of the central government had to be taken care of in order to supply the army. One might conceivably compute the yearly needs of the soldiers on the per capita, pre-war basis of one bushel or fifty pounds of salt. If there were 424,752 men in the confederate army,[4] striking an average for all the armies at any given time, then presumably that number of bushels was needed to meet the needs of the army for salt rations, for the salt for packing the soldiers' meat, and for salting hides for leather. However, we have learned that the monthly ration allowed about the middle of the war, was one and one-half pounds per month.[5] Since this basis of computation does not allow for the amount used to cure his meat or to preserve the hides for his leather, but, on the other hand, also does not allow for the many occasions when for one reason or another the full monthly quota could not be provided, perhaps it would not be desperately far from the mark to compute the average allowance for all purposes at thirty pounds per soldier a year, slightly more than the allowance computed by the Virginia legislature for the civilian population, as the soldiers were the first concern. Then our 424,752 bushels, first suggested, reduces to 254,851 bushels a year. We know that the government had a contract with the Virginia lessees, the Stuart, Buchanan Company for 22,000 bushels a month to April, 1862, and for a large indeterminate amount, but we have good reason to believe that this contract was not fulfilled. But we know from an authoritative report by the Board of Supervisors that Virginia, after she took over the works in July, 1863, up to December, 1864, delivered 144,335 bushels to the Confederate States government.[6] Since we know furthermore that it had a contract with at least two individual producers at the North Louisiana salines; since we know that it took over supervision of the New Iberia mines and produced from one pit for army needs; since we know that the Subsistence Bureau was an operator on the Florida coast, producing until disturbed by the union navy 400 bushels daily;[7] and since, finally, it would attend first of all to the soldier's needs, it seems entirely probable that this minimum war need for the soldier, despite all losses, was met on a subsistence basis at least. Indeed, no less an authority than Lucius Northrop, Commissary-General of Subsistence, is responsible for the following definite statement, dated January 25, 1865: "The supply of salt has always been sufficient, and the Virginia works were able to meet the demand for the army; * * *" This statement appears as a part of an important report submitted to the secretary of war just on the eve of the collapse of the confederacy.

It is further not unreasonable to believe that many, if in all likelihood not all, instances recorded when the soldier was in absolute destitution for salt, eating oysters, corn, and meat without salt, were due to the exigencies of a time of war, when a soldier frequently has lacked necessities, although there was abundance of the article in the country. This was probably often true after the great lack of this commodity had manifested itself in the fall of 1862, and after the government had frankly faced this question of salt supply as a serious problem. Certainly there is repeated evidence of huge stores of this article collected at various points for the army. The commissary of subsistence at Vicksburg reports, for instance, receiving on January 23, 1863, 104,293 pounds of salt to add to the 183,928 pounds already on hand.[8] According to his calculation that should have provided around 5,000,000 rations of salt so that it would seem that the besieged victims in Vicksburg, civilian as well as military, probably did not suffer for salt, whatever their other deprivations.

Probably the most direct way of examining the problem of the degree to which the actual needs of the civilians were met would be to study it by states.[9] For a few of the states we have an estimate of the amount of salt needed on a pre-war basis, reckoned by leading men of the given states, whose calculations are entitled to consideration. But possibly the better standard is that set by the special committee in Virginia which declared twenty pounds per capita the quantity deemed necessary, as there is no good reason to regard the civilian needs of Virginia as different from those of the other states.

In Louisiana it is impossible even roughly to compute the amount produced at the northern salines so as to add it to the rock salt output from the southern source in that state. The state was so early divided into union and confederate areas by the seizure of New Orleans that it is futile to query whether the salt obtainable were evenly distributed. Briefly catalogued, we know that Rayburn's Lick yielded around 1,000 bushels a day; Drake's more than 100 bushels a day, since that was the output of one firm only; Price's from 300-400 bushels; while there seems no way to estimate the output at King's or at Lake Bisteneau except to say that at the latter it was vastly greater than at any of the other salines, since 1,000 to 1,500 men were engaged in salt-making at this point.[10]

One may venture upon definitive figures for the rock salt from New Iberia, basing them upon the statement of the owner of the mine, who presumably would have kept figures of the production of each person or government functioning there, since his rental or toll was based on each bushel quarried. During the eleven and a half war months that the mine was in operation, from May 4, 1862, to April 17, 1863, when it was forcibly seized by the United States, about 1,360 bushels a day or 22,000,000 pounds in all were taken out.[11]

For Florida, fortunately, we have an estimate from so authoritative a source as the governor of the amount of salt needed to meet the needs of that state for the packing season.[12] He declared that "12,000 sacks of ordinary size", which would be 36,000 bushels, since the "ordinary" sack held three bushels, would be an "abundant supply". Although the amount produced on the coast of this state surpassed that of any other state,[13] except, of course, always Virginia, there was actual want for salt in the state, despite a daily production of more than 7,500 bushels for a portion of 1863.[14] This fact was due to the presence of tremendous numbers of salt-makers from other states so that the bulk of the product was exported and the governor was reduced to the necessity of suggesting, as has already been noted,[15] in order to keep some of the salt for the use of Floridians, a tax of one-tenth in kind. The tremendous destruction wrought there also makes it difficult to know how much was really available for consumption.

When the reader reaches Mississippi, he finds much more solid ground on which to stand, for, though figures are still far from complete, there are enough so that here one laments gaps whereas before, in Louisiana and Florida, one snatched at a few stray items. There is, to begin with, the 40,000 pounds of Louisiana salt which reached Vicksburg and was there stranded so that it required all the executive pressure which Governor Pettus could exert to pry it loose for distribution.[16] A report of the amount made at the Mississippi state works in Clarke County, Alabama, exists for the period up to January 1, 1865; and, then for the period January 1 to July 20, 1865. Since there must inevitably have been an almost complete cessation of salt-making with the close of the war in April, this record is to all intents and purposes complete and accurate. Furthermore, it will be remembered that Mississippi had no state works in Virginia or on the coast. The total amount manufactured by Mississippi during the period from 1863, when her works began to operate, to July 20, 1865, was 673,297 bushels. If to this amount be added the 77,958 bushels which she had received from private contractors and the 800 bushels (40,000 pounds) from Louisiana one has the sum total of state salt which this state was able to secure for the period of its activity, —762,055 bushels. This covers, however, it is necessary to point out, a period of a few weeks less than two years. For clarity let us say that Mississippi produced an average of 190,514 bushels a year,[17] reducing her production to the four-year basis of the war period. However, very far from this total was available for distribution to the civilian population, for many thousands of bushels had to be used to pay for supplies of one sort and another to keep the works going under the system of barter which prevailed. Indeed, as we have already learned, supplies were procured for salt which could have been had on no other terms; salt was exchanged for bacon, osnaburgs, shoes, corn, pork, lard, beef, and peas. The ratio established was three pounds of pork or four pounds of bacon for one of salt, but fifteen bushels of salt were required to buy one bushel of corn. It was used to pay for the labor of bricklayers on the furnaces, to pay for boarding mules, and to procure timber for the salt-warehouses.[18] It would appear, in fine, that only 580,711 bushels were distributed to the counties for citizen use under the four distributions which were held in this state. This would make the yearly average provided to Mississippi civilians by the state 145,177 bushels, again on a four-year basis. Since the census of 1860 gives this state a population of 791,305,[19] from which may be deducted 70,295 as representing the men subject to military service in 1860, the numeral arrived at, 721,010, roughly represents the civilian population of the state. The state was able, therefore, to provide about ten pounds per capita, assuming equal distribution, which is, of course, a false assumption. Furthermore, this calculation takes no account of private production and purchase. A neater way to estimate the state's efficiency as a provider of salt would be to say that on the basis of a quota of twenty pounds per capita, Mississippi was able to supply salt to about one-half of her population.

The paucity of available statistics for Alabama, which has such full financial records, as we shall see, is disappointing. Nowhere has the writer encountered complete figures as to the number of bushels produced for the state at the state reservation at the Upper Works nor the amount sent by Messrs. McClung and Jaques from Virginia. The only figures which have been found are in a report by D. C. Green, the quarter-master general, for the quarter ending March 31, 1863, that he had turned over to county judges and agents 269 barrels and 869 sacks of salt; and in the further record that for the quarter ending September 30, 1863, he had transferred to judges and agents 1,533 sacks or 4,599 bushels. Finally, we secure a little light from Woolsey's return for the quarter ending March 31, 1864, which shows 4,494½ bushels of salt made and 3,405 sold, or an average of 1,495 bushels a month, an amount which compares very favorably with the state output of Virginia.[20] The estimate of the total amount produced at the Alabama wells has been placed at 500,000 bushels a year.[21] The amount made at Bon Secour Bay near Mobile from sea-water amounted to 75 bushels a day [22] or 27,375 bushels a year, but was evidently made by individuals or for private sale so that there is no way of measuring the degree to which it helped to meet general needs through an effort at regulated supply. Large quantities must have been shipped by McClung and Jaques from Saltville, as is sufficiently attested by the large sums paid them, though exact figures elude us.[23]

In North Carolina, where officials computed the salt requirement at 18,000 bushels a month, we again meet definite and comprehensive figures. As early as the middle of August, 1862, the salt commissioner of that state, operating state works on the coast near Wilmington, was reported as manufacturing an average of 200 bushels a day, though the whole quantity made in that vicinity was about 800 bushels a day.[24] By September he was making 250 bushels per day till interrupted by the yellow fever epidemic, but the entire coast at this point, private and state works included, was yielding 2,000 bushels a day. By the beginning of the next year, Governor Vance reported 350 bushels of salt as the daily product from that source.[25] The nearest approach to a total from the coast works is that given by Salt Commissioner Daniel Worth for the year April 30, 1863, to April 30, 1864, which was 62,000 bushels, to which amount must be added 21,000 bushels made by J. Milton Worth during the period of service from December, 1861, to December 16, 1862.[26] This leaves a gap from December 16, 1862, to April 30, 1863, for which figures are not available, but during which period the production was certainly decidedly less than the 200 bushels a day recorded for August, 1862. As these definite figures from two independent sources then shows the production for just about two years (December, 1861-December, 1862; and April, 1863-April, 1864), it is probably fair to compute the average for two years as 114 bushels a day; and to draw the further conclusion that the earlier figures of salt production for this state are exaggerated,—do not, at least, represent constant, steady production. Certainly, it would not be safe to assume a higher average for the year, 1862, when there came the serious interruption of operations for more than a month due to the epidemic, nor for the six months from December, 1861, to June, 1862, when the works were getting under way and were further hampered by removals from Currituck Sound to Morehead City and then from Morehead City to Masonboro Sound.

By November 27, 1862, Woodfin had made about 36,000 bushels at the North Carolina furnaces in Virginia, and was able, so far as equipment was concerned, to reach 1,200 bushels a day. The contract with Stuart, Buchanan and Company called for brine for 300,000 bushels a year or about 1,000 bushels a day, but for many weeks after the agent was able to handle that amount of production, he was unable to get the brine.[27] Hence the only hope of even approximating to the actual production is to take the agent's statement of output as reported at intervals. By July 1, 1863, Woodfin had shipped 86,729 bushels in all, and had 20,000 bushels on hand, according to his own statement;[28] inasmuch as he had been operating since about October 1, 1862, it is possible to calculate his monthly production for the period of nine months at about 11,858 bushels, which amount may be a fair average for his capacity. This production at Saltville, added to Commissioner Worth's 3,420 bushels produced monthly at Wilmington sets the North Carolina contribution at 15,278 bushels monthly or about 510 bushels daily to meet the 18,000 bushel requirement for this state, a very favorable showing. Indeed, it would appear that North Carolina very nearly met her problem.

We are rarely fortunate in regard to the most important state of all, Virginia, for the precise figures compiled by the State Board of Supervisors in response to a senate resolution, calling for a "report in detail of the sale and delivery, for barter or otherwise, of all the salt manufactured and so delivered to the superintendent of the salt works of Virginia, manufactured and so sold under the existing laws," has been preserved. These figures cover, naturally, only the period from assumption of the works by Virginia in July, 1863, to the date of the submission of the report, December 31, 1864, but are indicative of the amount produced at the time when the wells were at the peak of production. The data were compiled from the monthly reports of the superintendent, exhibiting the amount of salt manufactured by him, the amount received or sent from the tenants, the amount distributed to the counties and towns of Virginia, the amount delivered to the confederate states, and the amount bartered for supplies.[29] There had been manufactured at the furnaces conducted by Virginia and paid her by tenants and contractors during this period of very nearly eighteen months 804,614 bushels of salt, of which amount 144,335.17 bushels had been delivered to the confederate government, and 590,264.25 bushels delivered to the Virginia counties. Here, as elsewhere, a considerable amount had been used for barter, 63,957.06 bushels, and some, 4,361.25 bushels, for negro hire.[30] Noting the amount delivered to the counties and reducing it to terms of a yearly basis, we obtain 393,510 bushels for twelve months to be allotted to a civilian population of 1,479,449 or [31] about thirteen pounds per person, after the requisite deduction is made for the Virginia soldiers. It is to be noted that the relatively small average of 1,078 bushels per day [32] has reference only to the amount produced by Virginia on state account and does not include the product of the furnaces of Georgia, Alabama, North Carolina, and Tennessee, nor that of private furnaces securing their brine from Buchanan, Stuart and Company by private arrangement. No record is complete which does not note the remarkable total of production from these wells as a whole. The daily output rose to 7,000 bushels before the close of the year 1862.[33]

The inadequacy of the supply in the fall of 1862 was strongly emphasized by the fact that the governor declared 1,800 bushels needed for the troops of the Virginia State Line. The fact that the Joint Committee on Salt in October, 1863, recommended that the distribution in Virginia be reduced from twenty pounds, as had been the practice, to ten pounds per person in order that the supply might be more generally distributed during the pork-packing season [34] reveals clearly that the state was not at that time meeting the needs.

For Georgia very full and detailed reports for an important period of the war are available in regard to the amount of salt delivered by Colonel Whitaker, commissary-general, and by J. R. Wikle, agent for the state at Saltville, on state account to an agent at Atlanta. Statistics are also available for the issues to the counties, covering the period of January, 1863, to April, 1864. The fact that distributions reached over half the counties, scattered through the entire state, argues that this depository at Atlanta was a general salt depot and that the record gives the full amount of salt handled on state account for this period. There was delivered to the state 139,044 bushels of salt; there was distributed to the counties only 63,990 bushels,[35] while 322 bushels were sold for the use of the state troops. In juxtaposition with these figures should be placed the pre-war consumption of 700,000 bushels a year, which on a reduced war basis of twenty pounds per person instead of fifty pounds would translate into 280,000 bushels. In a population of 1,057,286, from which a quota of 111,005 soldiers are withdrawn, one finds a civilian population of 946,281 to be provided for [36] which indicates an allotment of only about six pounds per person. It is further worth while to note that a very considerable proportion of this amount seems to have come from the works conducted in Virginia for Georgia; a statement from a committee, reporting in April, 1863, declared that the daily manufacture at Saltville on Georgia account was 1,500 bushels.[37] Hence, it is apparent that Georgia was far from meeting the needs of her citizens, and hence the constant concern of Governor Brown over the subject is explicable. Figures are, naturally, inclusive.

The production in the Trans-Mississippi region, which was not available to any extent east of the Mississippi after the middle of 1864, and which, therefore, must be thought of as contributing only to supply the local needs, should be briefly noted. No estimate of the production from the Arkansas springs has been encountered, other than the twenty bushels a day from a spring in Sevier county already noted. The yield from Texan springs indicates 240 bushels a day from Steen's Salines, 300 bushels from Brooks Salines, while the product from Grand Saline in Van Zandt County, the largest and the best, reached 3,000 bushels daily.[38]

Consideration of the cost of manufacturing salt also presents insuperable obstacles to final figures, but sufficient data exist to afford some conception of how the states of the confederacy poured out their treasure to meet this humble and ordinarily non-existent item in the budget.

The records are almost silent on the subject of the expenditures made by the confederate government in behalf of salt. A few stray items from the unpublished official manuscripts preserved in the war department may be noted. An advance of $100,000 on a contract for 300,000 bushels of salt with the Stuart, Buchanan Company on February 5, 1863, has already been commented on; a record of payments on freight charged to that same Virginia firm and evidently passed on to the Richmond government, reached a total of $637,091.76 for a period of ten months, from October, 1863, to July, 1864.[39]

It may be well to begin the study of the expenditures by the states for this one item with the bald statement of the total appropriation and then follow with some examination of the way in which those sums were distributed. Every state at varying dates and in varying amounts dipped into its treasury to meet the lack of salt. From first to last the states of the confederacy appropriated nearly $5,500,000 to this one object.[40]

The plan of studying expenditures state by state will be pursued again as the one most likely to yield what light can be thrown on this phase of the subject. Authoritative figures are difficult to ascertain, not because legislative appropriations are not clear, but because it is impossible to learn just what portions of those sums were refunded to the treasuries. Of his own authority, as an emergency measure necessitated by the war, Governor Brown authorized the commissary-general of Georgia to draw upon the funds of the state railroad up to $50,000 for the manufacturing costs and freights on salt, which it was planned should be replaced by the proceeds from the sale of the salt.[41] Indeed, it is difficult to see what else he could have done, for cash was immediately necessary to enable the agent at the Virginia salt works to get operations underway.[42] It is clear that the expense for the distribution of one-half bushel to each indigent family of a soldier, the cost of which was estimated at six dollars, was to be met out of the relief funds.[43] This, of course, did not in any way relieve the drain on the treasury but it did provide a proper legal source, which the governor could tap without special legislation.

A transcript from the Salt Account Book of the adjutant-general's office for the period from September, 1862, to March, 1863, would serve as a sample of the expenditures of this state to meet the salt problem. The largest items are those paid to the Georgia salt agent at Saltville, which were $10,000 for September, $30,000 for October, and $10,000 for January; and sums paid to Colonel May, obviously refunds to the railroad treasury on the $50,000 loan, $20,000 in October, and $8,500 in January.[44] The state spent $4,000 on freight during the period, while the other items were small sums, of only a couple of hundred dollars, probably salaries to various salt agents. However, they swelled the sum total paid out from state funds for salt for six months to $90,385.50. For purposes of a more exact comparison between the credit and debit sides of the ledger the figures for expenditures for salt from September to December are singled out and found to amount to $66,737;[45] for exactly the same period there seems to have been paid back from sales of the salt by the Inferior Courts, $55,606, almost five sixths of the sums expended, while it must not be forgotten that almost a third of the expenditure was made to repay an original loan, which over a long period of time might be regarded as a capital investment in the salt works. The receipts from the sale of salt are recorded for each year of the war and amount to the very considerable sum of $608,010.59;[46] unfortunately, the expenditures for salt during the whole period of the war do not appear in the same clear way.

The records of expenditures in behalf of salt in Alabama are painfully detailed, every penny being accounted for, from an item of $4.15 for sage and pepper, to the last postage stamp; from three awls and a punch, which cost $1.25, to an item of $15 for a coffin. Still it is difficult, without an expenditure of labor which hardly seems justified, to arrive at the total for the war period.[47] Some of the larger items may profitably be noted: Mr. Figh, lessee of the Lower Works, received in all $23,635 during the period 1861-1864; McGehee, the salt agent stationed at the works, received during the years 1862 and '63 the sum of $156,679.60; to the Alabama Salt Manufacturing Company for the same two years was paid $129,589.12; to McClung and Jaques for the same period, $33,064; to J. H. Speed, an agent for the state at Saltville, $20,300; to Snodgrass, a transportation agent, $10,020; to J. H. Bradford, still another special salt agent, $70,000; and Woolsey, who replaced McGehee late in the war, received $160,985.67 during a single year, 1864.[48] The sum total, therefore, expended by Alabama mounts merely for these items to well over half a million dollars. More illuminating, perhaps, are the comparative statements of sums expended in a given quarter juxtaposed with sums received for the sale of salt. For the six months from September 30, 1862, to March 30, 1863, Commissioner McGehee spent $70,249.66 and he received from the revenue on the salt and various sundries $73,362.45; during the first quarter of 1864 Commissioner Woolsey spent for the state $53,740.96 and he took in from salt sales and sales of property only $15,568.59.[49] It is conspicuous that on the basis of these figures, the state of Alabama spent far more than it received from the disposal of the salt. For one quarter, that ending September 30, 1863, it was remarked that the adjutant-general, who was still responsible for the salt production before the division of duties, expended $11,355.54, while he received from county agents $49,698.44 [50] but this large reimbursement may be accounted for by the fact that the planters were evidently laying in a supply of salt for the winter packing season. But, of course, no figures for a single quarter are conclusive; they should be balanced over a period of time.

Woolsey's property return for the quarter, ending March 31, 1864, shows a large and varied amount of property acquired by the state in pursuit of the salt industry.[51]

In Mississippi, as in Alabama, it is difficult not to lose the big totals in the maze of detailed items. Before proceeding to round sums, some of the particular items challenge interest. There occurs, for instance, the item of $250 for services of a surgeon and physician for two months at the Mississippi salt works in Clarke County;[52] there is the grindstone which Philips took with a salt well when he bought out a half-interest in a private well, which was paid for out of the Military Fund.[53] The sums exacted for freight and drayage would have been appalling even to public officials except in a time of war. The bill for shipping 57 barrels of salt from Iberia to Mississippi amounted to almost a thousand dollars.[54] The mere expenses of the agents whom Governor Pettus dispatched in his frantic search for salt in the summer and fall of 1862 contributed to swell the sum total. On the other hand, economy is manifested in unexpected places, probably because the given article was scarce. A case in point is that of a certain county agent who bound himself in writing to return to the general salt depot forty sacks in sixty days.[55] Furthermore, the items, which seem to comprise all the food bought to subsist the hands for her salt works from November, 1863, to May, 1864, are very modest.[56] There appears the same varied assortment of properties which any industry assembles, enlivened by amusing charges, such as the following: men were paid for patching the sacks; salt was bought to resalt pork; mules were sent out to board at forty cents a day;[57] stretchers suggest the accidents which we know occurred at the works; numerous whips and oxen remind of the eight-ox teams which moved the heavily laden wagons; lumber came to the works to build flat-boats for transportation on the Tombigbee. An item of nineteen dollars for hunting state mules for four days suggests a quality not usually associated with that animal. The sum of $336 was paid for advertising notices of runaway slaves;[58] in an abstract of articles lost at the state Salt Works, the agent conscientiously reported one old mule which died September 30, 1864, of the colic.[59] A grist-mill and saw-mill explain themselves. Tools would naturally be required, but the list of tongs, vices, sledge-hammers, saws, planes, chisels, shovels, locks, chains, rope, axes, and mules help to swell the investment of tax-payers' money. Negro labor was low—thirty dollars a month, whereas white workers, presumably in a supervisory capacity, commanded $125 a month. In face of such amassing of property there is little room for surprise that the state spent under the authorization of the act of January 1, 1863, which appropriated $500,000, $120,000 in the succeeding ten months.[60]

The sale of salt in Mississippi in the counties was based on the pound, rather than on the bushel as elsewhere, and on a sliding scale, for the price varied from thirteen to sixteen cents a pound. The cost to the state for the salt ranged from ten to sixteen cents, though the agent, Turner, computed the cost on one lot at thirty cents a pound.[61]

The North Carolina venture was unique in that it succeeded in paying its way, the price charged the counties being readjusted frequently so as to conform to the cost of manufacture. It rose from $3.50 a bushel in December, 1862,[62] to its high point, fourteen dollars late in 1864.[63] Commissioner Worth in January, 1863, declared that if he could proceed with the work without interruption, presumably selling at the current price of five dollars a bushel, he could in four months pay back all the money advanced from the treasury, which was then $61,-741.07.[64]

The capital investment in the works at Wilmington as computed by Worth on May 6, 1864, was estimated at $190,883.45, which sum included such heavy items as the Steamer, *J. R. Grist* [65] and three flat-boats, worth $6,800, together with 52 horses and mules, which with their wagons and harness he put down at $18,200. He reported only $100,000 drawn from the treasury. In addition he recorded large items, evidently purchased from profits, such as salt pans, costing $30,000, and fixtures worth $40,000, so that he shows a balance in favor of the works of $794,595. He declared that if the prices charged by the state were compared with the market price of salt,[66] the state works at Wilmington had effected a saving to the people of North Carolina of $697,500.

In Virginia, where the state preempted a large portion of the salt works, the reader is dealing with "big business" in a very real sense. The annual rental called for payments to Stuart, Buchanan and Company of $773,000, while the rental for the Charles Scott furnaces amounted to $549,177.50.[67] The investment of the state in the enterprise, was reckoned by the superintendent on March 1, 1865, counting eighty horses and mules, 25 wagons, 120 beeves, and a large store of provisions, at $1,123,-000. But against these assets were set liabilities to the original lessees, and sums due for negro hire and freight of $1,338,000.[68] While the amount of money handled was large, evidently it was not lucrative, as this financial statement would seem to indicate that the liabilities absorbed the assets.

Nothing more can be stated with regard to the expenditure in Florida and Arkansas for this commodity beyond the bare fact that $21,000 was expended in behalf of salt, according to the governor of the former,[69] and that $500,-000 was appropriated in the latter in two installments—$200,000 allocated from the general relief fund on one occasion, and $300,000 on another occasion.[70]

CHAPTER XIV

# CONCLUSIONS

THE FIRST conclusion which obtrudes itself upon the thought of the reader is the transcendent importance which the most insignificant commodity can assume in time of war. All articles of food become factors of the first importance the moment a blockade threatens to shut off any portion of a belligerent's territory from its accustomed source of supply. In the four-year contest between the federal union and the confederacy, salt was such an element of prime importance in the diet. With our present methods of refrigeration and preservation of fresh foods of all kinds, salt has lost its earlier importance, and this story will never need to be rewritten for another war. But the question has merely changed its form: in the Civil War, the scarcity of salt was a vital question; in the World War lack of wheat for the continental countries was the serious factor, which, by the same token, caused the loaf of white bread almost to vanish from the American table, as we voluntarily assumed the eating of corn bread, rather than impose it upon the European peoples, who were already suffering grievously under war conditions.

The tremendous amount of salt consumed instantly challenges the attention and emphasizes the changes possible in dietary questions within scarcely three-quarters of a century. The meat of the winter season, which was apparently confined for most of the southern population to bacon, smoked ham, and beef pickled in brine, would probably be disdained by the descendants of the planter class and has probably vanished largely from the tables of the humbler folk. For a population of nine million people to be asked today to consume nine million bushels of salt annually would doubtless be felt a grievance.

It will probably occur to the reader as striking, that despite the most persistent search, despite enthusiastic boring to locate subterranean brines, despite scientific testing of weak brines, no important new sources of salines were discovered, with the one exception of the mine of rock salt at New Iberia. That no new wells of real salinity were located is not surprising in view of the fact that we are dealing with a country thoroughly explored. But it is passing strange that with a group of four large rock salt masses lying below the surface in southern Louisiana close to Petite Anse, and with large salt domes in Texas, no persistent search was undertaken which would have located some, at least, of these invaluable treasures. The explanation probably is to be found in the fact that after the capture of Avery's Island by the union forces, the hold of the confederacy upon the Trans-Mississippi region was too weak, confederate energies too completely engaged, to leave any strength to embark on a hunt for salt. Probably, furthermore, the scientific information which suggested to a Thomassy or a Hilgard the probability of the existence of salt masses below the surface was too little disseminated to prompt any search.

The reader cannot fail, however, to observe with wonder the really remarkable piece of industrial enterprise in salt manufacturing which must always remain to the credit of the confederacy and of the individual states which composed it. It should be noted that salt manufacture, along with the production of arms and clothing, constituted about the only manufactures upon which the confederacy was able to embark. The central and state establishments for making salt were larger, their output greater, and their administration far better than even a sanguine confederate would have been justified in predicting. This is the more remarkable in that the industry, except at Saltville, had to be built up from the ground, and under war conditions. The persistence of the southerners, manifested at St. Andrew's Bay, on the Florida coast, in rebuilding the works so constantly destroyed by the federal fleet, is only equalled by the speed with which the confederate government re-erected the furnaces and recast iron kettles. It is difficult to explain whence the Richmond authorities assembled the money, materials, and labor for this really astonishing feat.

One must also express surprise and admiration for the energy poured out on this work by the governors and legislators of the states, almost without exception. It is true that one can understand the captiousness of those critics of gubernatorial policy who held that the same energy might better have been directed toward securing arms to win battles or toward breaking the blockade in order to permit the entry of foreign salt, but the governor's zeal is explicable. Salt was rightly recognized as absolutely fundamental for the food supply and the assurance of food supplies is a first rule of warfare. Though the word "morale" which was so incessantly on the lips of our leaders during the late war seems not yet to have entered the working vocabulary of the war governors, the thought of it was clearly present and a factor in their thinking, for more than once they speak of the danger of mutiny if a salt supply could not be provided.

The study in pounds and bushels pursued in the preceding chapter, demonstrates clearly that the indefatigable efforts of officers and executives were not crowned with the success which they merited. While the central government did secure enough stores of salt to meet the needs of the soldiers, on a war basis at least, the states do not seem to have been able to do equally well for the civilian population. The states which were unfavorably situated by virtue of having no available saline resources within their own limits, such as Georgia and Mississippi, suffered sadly at times for the want of salt, while in all there was a sufficient lack to make it an object of anxious solicitude. The stringency became acute each year as the packing season approached. The fact that this was true despite strenuous efforts was due, in part doubtless, to faulty transportation, but also, possibly in larger measure, to the constant destruction of the salt produced before it could be distributed.

The modern reader cannot fail to realize that certain questions which seemed to interpose insuperable obstacles to the rapid production and transportation of salt would be readily solved today, thus emphasizing the changes wrought by three-quarters of a century. Pipe lines would promptly carry the brine, not five miles into the woods to the source of the fuel supply, as was proposed, but thirty, fifty, a hundred miles to points beyond the congested section of the railroad branch from Bristol to Saltville. In passing, furthermore, we must not forget that the modern industrialist is not now depending on the costly, bulky wood for his fuel.

The solution of this one item, the transportation of the brine or salt, would have prevented some of the friction between the states within the confederacy, which was certainly no help to the cause. The ill feeling engendered by delays in the removal of the salt and fomented by many caustic messages, by bitterness on the part of the salt agents resident at Saltville, by retaliatory measures, urged and occasionally adopted, did not promote the hearty co-operation indispensable for a successful prosecution of the war.

As is always true, the heaviest burden of the suffering fell on those least able to bear it—the poor. The tale of that element of the population which was least able to secure the precious crystals to put up the carefully treasured two or three hogs is eloquently told in the badly written, half intelligible appeals which poured into the governor's letter files—pathetic calls which remind one today of a child's cries to its parent for help. There were times when the stringency in the article of salt alone brought portions of the population very close to the point of desperation and revolt. When respectable citizens move to conduct a raid for a given commodity, the bonds of orderly government are clearly menaced.

The price paid by the confederate states for their salt was tremendous, almost overwhelming to the modern reader. If one computes merely the lump sums appropriated by the various state legislatures—and one need have no fear but that they were duly expended—one arrives at the striking figure of about $5,500,000 spent for salt from public funds. Some, though not by any means all, of this sum was refunded to the treasury from sales of the product, for it was expressly stipulated by law in most of the states, that it should be distributed gratis to the families of the indigent or deceased soldiers. Some conception of the small proportion refunded is afforded by the following data furnished by Governor Pettus to the Mississippi legislature on November 3, 1863: he recites that $120,602 of a sum appropriated on January 29, 1862, had been expended on account of salt, and that $12,589.67 found its way back to the treasury through sales of salt. The modern citizen will naturally regard these large public funds as diverted strangely from other enterprises more legitimately within the realm of public works. He will also look askance at the waste of tax-payers' money, owing to such actions as that of the Norman Company in 1863, which, after accepting advances in money and provisions, asked release from the contract with Mississippi on plea of utter inability to fulfill it; and at such unbusiness-like methods as the sale of wet salt so that it lost one-fifth of its weight in transportation over relatively short distances, as has been noted. Furthermore, a good business man could not fail to note the large sums of money tied up in the state salt establishments, $200,000 at the relatively small works of the North Carolina coast, to cite a single instance, which fact was bound to result in loss and waste. Serving a temporary war purpose, we may be certain that the property at the other works was scattered and lost, or disposed of at a ruinous sacrifice, as we know happened at the Mississippi works in Clarke County when the commissioner found himself in the situation of being forced to liquidate the business. He certified that he tried to dispose of the property to "the best advantage," but the steam grist mill and saw mill had been burned by some unknown person and much of the property was "in a damaged condition and realized prices accordingly." [1]

This commonplace necessity determined, as has been abundantly described, military manoeuvres, so that the wells on the Goose Creek, in the Kanawha Valley, and in the mountains of southwestern Virginia became the objectives of military attack. The destruction of stores of salt or the machinery of its production became a feat of equal value with the gaining of a battle and was esteemed by the north worth a costly military expedition. On the one side, the confederates boasted that the gaining of the Goose Creek Salt Works in Kentucky was worth more to them than all the other spoils acquired in that state. On the other hand, the leaders of federal raids rejoiced that the breaking up of some salt works of the Richmond government near Pensacola injured the south more than the loss of a great battle, or of Charleston. The *New York Herald* moreover hailed the destruction of the works on St. Joseph's Bay as a greater blow to the "rebel cause than if we had captured 20,000 of their troops," [2] because it was felt throughout Georgia and Florida, where the inhabitants depended on these works for their winter's store of salt. Attacks on these works, whether by military force directed against Saltville, or by vessels of the blockading fleet against the shores of the Gulf coast, were fully recognized by union commanders as among the most easily accomplished and disastrous blows which could be struck at the confederates.

It would appear almost an act of supererogation to labor the point of one of the most tragic aspects of war—the dreadful, futile waste. The harassing of the industry at St. Andrew's Bay alone cost the confederacy over six million dollars, a fact so striking that the writer may be forgiven for repeating it. Add to that the loss by destruction of many other points along the coast, and the destruction wrought in Kentucky, Saltville, and many minor points noted in the *Official War* and *Naval Records,* and the sum mounts to many millions of dollars. Some meat supplies must have been lost because of the lack of salt to cure them. Frequently, the destruction dictated by the necessities of war appears wanton, as when the salt was mixed with sand or mud merely to render it useless, or was seized for the moral effect of dealing a serious blow to the foe. It is easy to imagine the mingled feelings of rage and despair with which a southerner saw northern soldiers callously pour out the white crystals, which were an article of such anxious solicitude to him, on the beach or into a stream.

The loss touched everything connected with the industry: the machinery for operating the works, the furnaces, the finished products, the sacks or barrels, nine thousand sacks alone being burned in the Saltville raid, the wood gathered for fuel, the hides from slaughtered beeves, and the account books.[3] Frequently, the confederates themselves destroyed supplies to prevent their falling into the hands of the enemy. On the debit side also must be entered the seizure of negroes, sometimes to the number of forty or fifty, and of teams, which had been hauling wood or supplies, and which were pursued by the unionists for many miles before they were overtaken.[4]

The ease with which the confederates could repair the works, and, hence, the great difficulty of dealing the industry a crushing blow constitutes one of the surprises in a study of the subject. Again and again the union commander writes with satisfaction that so thorough had been the work of demolition, it would be long before work could be resumed if, indeed, it would ever be possible; in a few days he was proved wrong, and the industry dared again to raise its head. The long-planned raid on the Virginia works at the close of 1864, which seemed at first blush so effective, really struck a very slight blow to the salt stores of the south; the North Carolina agent reported that the damage to the salt of that state would not exceed 1,000 bushels.[5]

However, in the end, here as elsewhere, superior northern resources for continuing the fight against salt-making had to tell, and ultimately the union navy virtually harried the industry from the coast. The effectiveness of the blockade was also demonstrated, in that it prevented the fulfillment of contracts for the importation of salt and frequently strewed the salt cargo of blockade runners along the beach.

Again, the reader is impressed with the loss of energy and time by a reversion to archaic methods. Planters resorted to the methods of their colonial forbears in sacrificing several weeks of each year to visit the salt-licks or sea-shore and laboriously to secure by the wasteful methods of individual rude furnaces and kettles the year's supply of salt, which before the war had been procurable for a few cents a pound. That same time and energy would, except for war conditions, have gone toward developing the regular crops, and, more productively, toward improvement of their plantations.

As always, war problems moved in a circle. Salt was indispensable and must be produced, but its production constituted an appreciable drain on the economic resources of the section in which it was produced. In Florida and in southern Alabama, for instance, the food consumed by the salt-makers, pouring in from other states, and the forage consumed by the horses and mules helped rapidly to exhaust the already rapidly dwindling resources of those states. Commissaries had to function to support the regular laborers at the salt works. The greedy visitors also, according to Governor Milton, injured the credit of the state by refusing to accept the treasury notes of Florida in all their fiscal transactions.[6] Wagons, horses, and mules, sometimes sadly needed by the state and counties for military purposes, were diverted for the use of the salt-makers. The use of fuel for salt-making was no exception to the universal law of extravagance and waste in time of war, for a swath was cut through the timber and wooded areas near the salt works, with the single exception of Virginia, where the legislature wisely decreed that one-fifth of the timber must be left standing.

Finally, the rough, archaic, and, hence, uneconomic methods of war-time manufacture could not and did not form the basis for a permanent post-bellum industry in salt in the south. It is conspicuous that the south reverted after the war to its old markets, importing large quantities from abroad.[7] The great centers for the industry, New Iberia and Saltville, continued for years to be two of the important sources for salt, but this was due to their natural advantages, which had, in the case of the Virginia works, been recognized before the war, and not at all to war stimulus. Mistakes were made in the initial exploitation of the mine, in leaving too thin a roof, through which the water percolated, and in spending money for dredging bayous for a system of transportation. Some years elapsed before it was even possible to command capital for the development of the mine on a commercial basis. On the hill to the north of the main shaft the effects of dissolving out salt from a mass overlaid by unconsolidated rock may be seen. An enormous funnel-shaped hole in the ground marks the place where several thousand tons of salt were obtained by the common method of pumping brine and evaporating it for salt. By this method that portion of the island underlaid by the salt mass could literally have been melted into the sea. Salt is no longer produced at Saltville, as the works located there are devoted to producing alkali by the Solway process. All production from the weak brines, given a brief renaissance by war conditions, soon ceased in the face of competition with salt derived by cheaper methods. The persistent efforts of M. Thomassy to advance the production of salt by solar evaporation were doomed to failure. This method of production did not secure a real hold in the

United States for years and then it was in Utah and California, not in the region of the old confederacy.

The fact that salt could become a major problem to the confederacy reveals strikingly the industrial backwardness of the south, its complete dependence on outside sources for primary needs and emphasizes that fact as the most serious of its disadvantages in the unequal struggle. The failure of the confederacy, though predictable from the start, was immediately attributable to errors of judgment in not anticipating and justly estimating its inability to supply certain indispensable necessities. It had not one organized industry to produce salt from the sea water which laved its shores so generously at a time when Europe was utilizing widely that cheap method of salt production. In a very real sense it may be claimed that by diverting men, materials, and capital from the first objective of war, winning battles, the lack of salt was a contributing factor to the outcome of the War between the States.

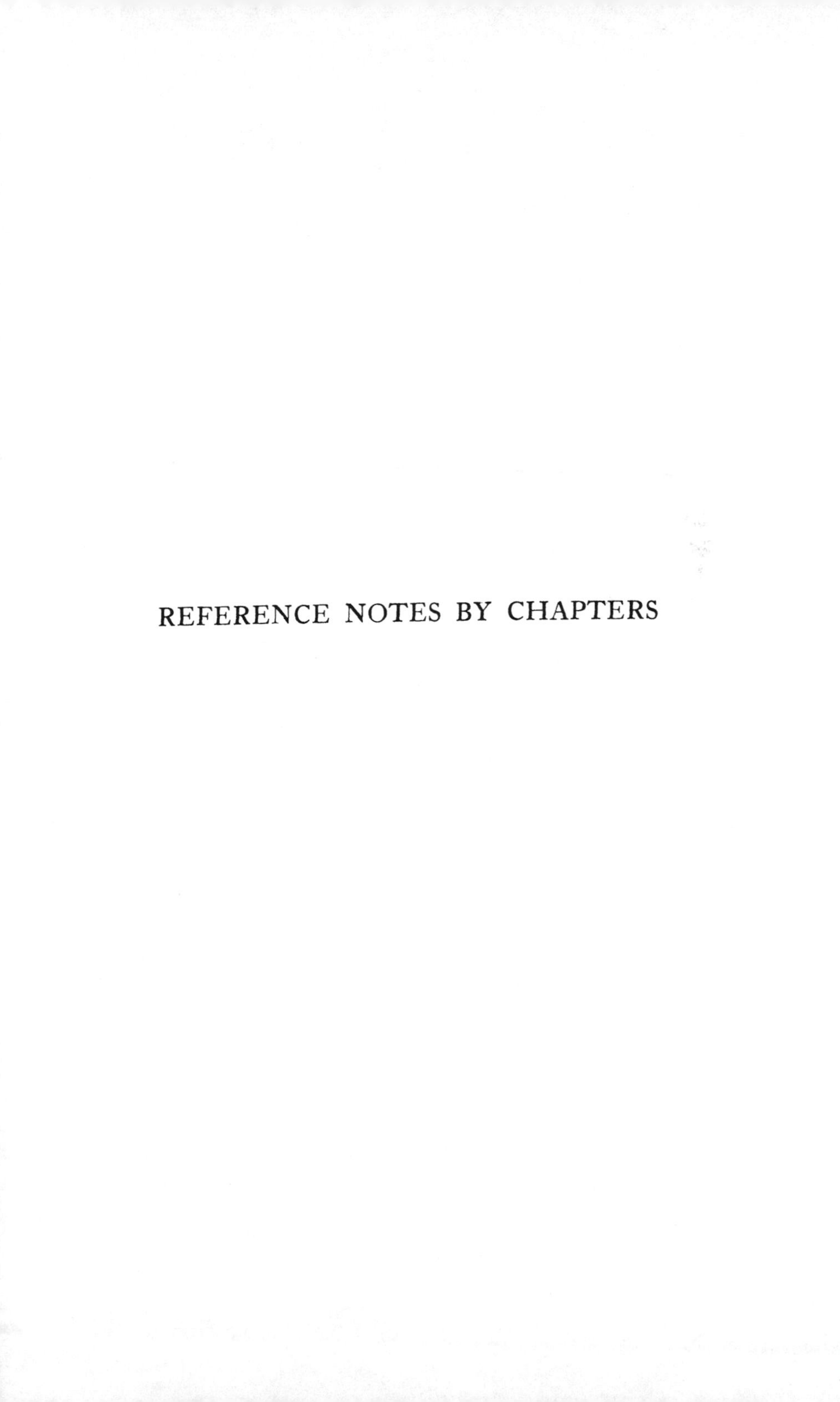

# REFERENCE NOTES BY CHAPTERS

# NOTES

## CHAPTER I

1. See the account of how the soldiers revolted against oysters without salt, this book, pp. 47-48; and of how roasting ears palled upon the palate. Booth, *Personal Reminiscences,* p. 71.

2. *Official War Records,* series I, volume XIX, part 2, p. 709.

3. "Bring only sugar, coffee, flour, and double rations of salt, driving the beef on the hoof." *Ibid.,* volume XII, part 3, p. 551.

"Bring all the supplies you can possibly carry, omitting every article of the ration, except coffee, sugar, salt, and hard bread. Accumulate beef cattle—a considerable herd." *Ibid.,* p. 441.

These instances happen to be drawn from the union army records, but the condition was applicable to the confederate forces.

Fresh pork was felt to be unwholesome. "This army is suffering from the use of fresh pork." *Ibid.,* volume XXIII, part 2, p. 625.

4. *Ibid.,* series IV, volume I, p. 1010. The writer has found the number of pounds to the bushel variously computed, ranging from fifty to seventy pounds, just as the number of bushels in a sack varied. She finds fifty pounds to the bushel and three bushels to the sack the ratio most widely used.

5. *The American Annual Cyclopedia,* 1864, p. 32. In 1862 the salt ration for the army was two to three quarts (four to six pounds) to every hundred rations. *O.W.R.,* series I, volume X, part 2, pp. 530-31; volume XVII, part 2, p. 638; in 1863 it was four and a half pounds to every hundred rations. *Ibid.,* volume XXIV, part 3, p. 648. The report of the secretary of war shows about the same estimate. Report of Secretary of War, January 3, 1863—November 3, 1864, Appendix, p. 32.

6. The allowance cited here prevailed in Richmond. The war clerk, Jones, writes in his diary on November 7, 1862, "Yesterday I received from the agent of the City Council fourteen pounds of salt, having 7 persons in my family, including the servant. One pound to each member per month is allowed at 5¢ per pound." *A Rebel War Clerk's Diary,* volume I, pp. 183-84. The estimate of thirty pounds per capita was for Virginia. "Minority Report of the Joint Committee on Salt", *Documents of Virginia,* 1863-64, Document No. 28, p. 7.

7. *On the Rock Salt Deposit of Petit Anse, Louisiana Rock Salt Company.* (Report of the American Bureau of Mines), pp. 24-25.

8. Thomassy, *Géologie pratique de la Louisiane,* pp. 187-89. Thomassy had made four or five tours of inspection of the Mississippi delta. The book cited, was published by American subscribers in New Orleans and also in Paris, (the first geological survey of the state). This interesting

character appears again in our story. He thought the failure to manufacture salt in the United States was due to defects in our mode of production of salt.

9. The writer arrived at this estimate by computing the civilian consumption at one pound a month per person, and the consumption of the soldiers, estimated at an average of 424,752 at any given time in all armies, at one and a half-pounds a month, which may be little more than a guess. It is necessary to explain that Livermore's figures and estimates for the number of soldiers are accepted. He establishes from the confederate tables of the numbers present and absent on given dates (*Numbers and Losses in the Civil War of America,* p. 47) the probable size of the confederate army at various periods of the war so that it is possible to compute the average number through the entire period at 391,845. To this figure it is necessary to add the men in irregular service. He places this number at 98,720, serving an average term of sixteen months each. It is possible to add one-third of this number (since the entire duration of the war was four years), to the above number and arrive at 424,752 as indicating roughly the average number of soldiers under arms at any one time during the war. It is not regarded as necessary for this rough estimate to calculate the deduction which should strictly be made for loss by death, illness, or desertion.

In adopting an average consumption of only one and one-half pounds per soldier per month, the writer is conscious that she is accepting the lowest of the figures available from the *War Records,* that for 1864, but she felt that this figure would probably come nearest being correct, since even in the earlier years the soldier often failed to receive his full ration, and during the last lean year fell almost constantly below his meagre ration.

10. Northrop, commissary-general, calculated that they would secure only about one-third of the requisite amount from the confederate states, Pollard, *The Lost Cause,* p. 484.

11. "Clements, who had been eating raw beef tongue, and it very salty and dry, was nearly dead for water." McMorries, *History of the First Regiment Alabama Volunteer Infantry,* p. 104.

12. Hague, *A Blockaded Family,* p. 24.

13. J. A. Worth to a brother, *The Correspondence of Jonathan Worth,* volume I, p. 225. So universal was this process of salting meat for the winter supply that but one use for salt was suggested during the packing season. This same brother had written six weeks earlier, "I had sent forward the salt some days ago. I suppose it is used by this time on your pork." *Ibid.,* p. 209.

14. This rule is cited as that of the experts in the business by a special committee of the Virginia legislature. "Report of the Joint Committee Appointed to Visit Confederate Authorities in Regard to Supply of Salt", *Virginia Documents,* 1861-62, Document No. 7, p. 7.

15. *O.W.R.,* series IV, volume II, p. 916. Note also the comment of Jonathan Worth to his brother, December 30, 1861, "An immense amount of salt must be ready for the fishing season in the spring." *The Correspondence of Jonathan Worth,* volume I, p. 162.

16. *O.W.R.,* series I, volume XXX, part 3, p. 43; volume XVII, part 2, p. 155.

17. It is a well known fact that animals will leave the best unsalted hay for that of inferior quality moistened with brine.

18. Hague, *op. cit.*, pp. 33, 42.

19. Evidence that salt was used on poor land is found in Clay-Clopton's diary, *A Belle of the Fifties*, pp. 223-24. Even such an eminent scientist as Thomassy held that view as is sufficiently proved by his articles in *De Bow's Review*.

20. The scorbutic condition induced by excess of salt modified the course of every disease, poisoned every wound, and lay at the foundation of dysenteries, we are told. No body of troops could be confined to the confederate ration, according to the best medical opinion of the day, without suffering materially in health and without manifesting scurvy. Stevenson, *The Southern Side or Andersonville Prison*, p. 74.

The writer understands that scurvy has since 1912 been attributed to the absence of vitamine C.

## CHAPTER II

1. Rock salt was at the outbreak of the Civil War being mined nowhere in the United States. Although the deposit in Virginia was recognized as rock salt after 1840, it was not mined, and the salt mine in Louisiana was discovered only in 1862. It was first learned that the salt wells in New York had their source from a bed of rock salt in 1865 after the close of the war.

2. *Official War Records*, series I, volume XII, part I, pp. 1151, 1152.

3. Laidley, *History of Charleston and Kanawha County, West Virginia, and Representative Citizens*, pp. 123-25.

By September, 1862, the Kanawha works were producing but little salt, owing to the lack of coal to run the furnaces. This lack of fuel was due in turn to the absence of blasting powder. This fact was presented to Secretary of War Randolph in September of that year. *O.W.R.*, volume XIX, part 2, p. 611.

For the story of the discovery of the wells see Patten, *The Natural Resources of the United States*, p. 310.

There were a few other places in West Virginia where salt was made—at Bulltown on the Little Kanawha and at the Mercer Wells in Monroe County. See *O.W.R.*, series I, volume LI, part 2, pp. 270, 286.

4. Owen, *History of Alabama and Dictionary of Alabama Biography*, volume II, pp. 1225-26; Ball, *A Glance at the Great South East or Clarke County, Alabama, and its Surroundings*, p. 645.

5. *Ibid.;* see also the last report of Z. A. Philips, Salt Commissioner for Mississippi, January-May, 1865, series G, No. 201, Mississippi Archives.

Even now the source of the brine in Clarke County is not definitely ascertained. It is thought that the water, migrating through the porous strata, dissolves the salt, either from beds of solid salt, or from connate water, rising through cracks and fissures in the formations by hydrostatic pressure. Barksdale, "Possible Salt Deposits in the Vicinity of the Jackson Fault, Alabama," Circular No. 10 of the *Geological Survey of Alabama*, p. II.

For the location of the springs see map.

6. *The South in the Building of the Nation,* volume V, p. 297. The name derives, of course, from the fact that wild animals resorted there to lick the salt which was deposited on the surface of the earth.

7. It must be understood that the word "bayou" was used interchangeably with river and does not always in local parlance mean a cut-off lake.

8. See extract of de Bienville's *Journal.* The passage referred to may be conveniently found in King, *Sieur de Bienville,* pp. 100, 102.

9. Veatch, "The Salines of North Louisiana." No. 2, in a *Report on the Geology of Louisiana,* pp. 51-57, 83. See map.

10. *Ibid.,* pp. 55-56.

11. *Ibid.,* pp. 57-59; Hilgard, *Supplementary and Final Report of a Geological Reconnoissance of the State of Louisiana,* pp. 31-32.

12. *Ibid.,* pp. 28-29; Veatch, *op. cit.,* pp. 71-73.

13. Veatch, *op. cit.,* pp. 65-67; Hilgard, *op. cit.,* p. 29.

14. Veatch, *op. cit.,* pp. 76-77; Hilgard, *op. cit.,* pp. 28-29.

15. Veatch, *op. cit.,* pp. 83-84. Though large scale production ceased with the war, an old negro continued a little salt trade until his death in 1892 at an old well in Crane Lake, known in his honor as Old Dan Bryan's Well, *Ibid.,* p. 86.

16. *Ibid.,* pp. 90-92.

17. Cedar Lick near Winfield and a spring a half mile above the mouth of Bayou Negreet are cases in point. *Ibid.,* 90, 92.

18. Owing to the great value of the salt wells, the dividing line between the two counties, when Smyth was formed from Washington County in part, was run so as to throw equal values of the great mineral resource into each of the counties, Maury, *Physical Survey of Virginia,* p. 106. (Preliminary Report). See map.

19. Phalen, *Salt Resources of the United States,* pp. 85-88.

20. Preston, *Historical Sketches and Reminiscences of an Octogenarian,* pp. 38-43. The story has been told that the lick was almost lost to the Campbell family very early, for Charles Campbell, the patentee, was arrested for non-payment of a small amount of taxes. He wrote his wife, directing her to sell the salt lick tract in order to rescue him from jail. She promptly spun into yarn many hanks of flax which were at hand for spinning, took it to market, sold it for more than the small amount of the taxes, and took her husband home to the salt licks. *Ibid.,* p. 50.

21. In 1781 Thomas Jefferson in his *Notes on the State of Virginia,* p. 51, mentioned the occurrence of salt brine in the Holston Valley, but the fact that the brine came from rock salt was not discovered until 1840, when a shaft was sunk which struck a bed of rock salt at a depth of 210 feet. It is amusing to record the disappointment of the operators; sinking a shaft for salt water and encountering only a dry shaft, they bemoaned their wasted money. Watson, *Mineral Resources of Virginia,* pp. 211-12.

22. William Campbell was the son of Charles Campbell, the original patentee, who had died in 1767.

23. Preston, *op. cit.,* pp. 53-54.

24. *Ibid.,* p. 55.

25. His well was located at the head of the valley so that surface

drainage did not weaken the water. It was so saturated with salt that it weighed nine pounds per gallon.

26. The statement of the amount of salt produced as early as 1805 is based on Jedidiah Morse's *Geography,* printed in that year and cited by Preston, *op. cit.*, p. 57. The earliest copy of Morse's book which the writer has been able to locate is *Geography Made Easy, An Abridgment of the American Universal Geography,* (Boston, 1806). This edition makes no mention of the salt resources of Virginia.

27. The terms of the lease are given in detail in Preston, *op. cit.*, pp. 58-61.

28. The fact that they owned the Preston estate is established by a letter from the firm to Governor Clark of North Carolina under date of June 15, 1862. Letter Book, 1861-62, Governor Clark, North Carolina Historical Commission, p. 344.

This fact seems to be so little known that it appears desirable to quote the passage referred to: "We are the lessees of the Preston and King's Saltworks' estate in Smyth and Washington Counties of this State, and we have in the last few days become the owners of the Preston Estate. This estate contains a flat of 200 or 300 acres and in different parts of it wells have been sunk for salt water, and we believe without exception the salt water was found within 250 feet of the surface."

29. Switzler, Report on the Internal Commerce of the United States, Appendix on Virginia, found in *House Executive Documents,* 49 Congress, 2 Session, volume 7, part 2, p. 149.

30. Compare the analysis given in Watson, *Mineral Resources of Virginia,* p. 212, and Switzler, *op. cit.* They differ slightly.

| | Watson | Switzler |
|---|---|---|
| sodium chloride | 99.084 | 98.39 |
| calcium chloride | traces | |
| calcium sulphate | 0.476 | 1.22 (sulphate of lime) |
| iron | 0.476 | traces |
| magnesium sulphate | | 0.39 |

31. *The American Annual Cyclopedia,* 1862, p. 12; *O.W.R.*, series I, volume LIII, p. 966; volume XLI, part I, p. 227. General Hindman reported to Richmond from that state on June 9, 1862, that he could produce salt in large quantities. *Ibid.*, volume XIII, p. 833.

32. Owen, *Second Report of a Geological Reconnoissance of the Middle and Southern Counties of Arkansas,* pp. 112, 115.

33. The map is entitled *Colton's Railroad and Township Map of Arkansas,* compiled by D. F. Shall. A photostatic negative of the map exists in the Arkansas Geological Survey. It appears that Shall, wishing to produce a map of the state which should show the salt springs so granted by congress sought information from his friend, Gordon Peay. The latter took his data from the plat on file in the State Land Office. The seventeen springs which he enumerates are located in what are now Johnson, Franklin, Cleburne, Pope, Hot Springs, Little River, Clark, and Sevier Counties.

It is a reflection of the time that the letter was addressed simply "City of New York", was folded and sealed without an envelope, and bore no postage. The amount paid by the sender was merely marked in one corner, fifty cents. *The Arkansas Gazette,* January 25, 1931. See map.

34. *O.W.R.,* series IV, volume I, p. 324; series I, volume XIII, p. 761.

35. Phillips, "The Mineral Resources of Texas," *Bulletin of the University of Texas,* No. 29, p. 218, Buckley, *A Preliminary Report of the Geological and Agricultural Survey of Texas,* pp. 19-20. Water from the spring last mentioned boiled down in about the proportion of one bushel of salt to twenty of brine. This salt supplied a large section of country.

36. Evidence that the coast was utilized for salt making even in Texas, on the Laguna de la Madre, is to be found in *O.W.R.,* series I, volume IX, p. 624.

37. Salt-making had been conducted on a commercial scale at St. Andrew's Bay, Florida, long before the war. The first salt company had been incorporated in the Territory of Florida in 1829, the second in 1833.

38. An excerpt from the *Natchez Daily Courier* of September 10, 1861, is illuminating as to how the people felt concerning the Texas supply: "There should not be such a hue and cry about the want of salt. There are lakes in Texas (not lately discovered, as some of our exchanges state,) which will yield almost enough of the article for the consumption of the Southern States. The only difficulty now is the transportation of it. Had the citizens of New Orleans come as promptly up to the work on the Texas and New Orleans Railroad as did the people of Texas, we would now have railroad communication with that state. As it is, there is a space of one hundred and twenty-five miles, open between Orange, the eastern terminus of the Texas portion of the road, and New Iberia, the western terminus of the Louisiana portion." See also *The New Orleans Daily Crescent* of December 2, 1861, where the writer hoped for salt from the Texan lagoons at twenty to twenty-five cents a bushel, if means of transportation were available; and also *The Clarke County Democrat,* August 12, 1861.

39. The geologist Harris speaks of such "dry salt lakes" of the Trans-Pecos region in western Texas, covering about forty-five acres, and of many salty streams in north-western Texas, along the banks of which fairly thick crusts of salt formed. *Geological Survey of Louisiana,* Bulletin No. 7, p. 100. For salt lakes near Corpus Christi see *The Weekly Raleigh Register,* October 2, 1861; Dumble, *Second Annual Report of Geological Survey of Texas,* pp. 44-48, 222-24.

40. *Session Laws of Texas,* November-January, 1861-2, Joint Resolution, chapter VIII.

41. The exact yield was as follows:

| | |
|---|---|
| Virginia (including the Kanawha works) | 1,900,000 bushels |
| Kentucky | 250,000 " |
| Florida | 100,000 " |
| Texas | 25,000 " |

The yield from Massachusetts, Michigan, and Illinois was as yet negligible, 5,000 bushels annually. *On the Rock Salt Deposit of Petit Anse,* p. 25.

The foreign importations amounted to 17,165,000 bushels a year.

42. This brief account of the early history of the well and of the discovery of the rock salt follows in the main Latham, *Black and White,* pp.

187-88. The author of that work talked personally with Judge Avery just after the war. I have also consulted the Report of the American Bureau of Mines, *On the Rock Salt Deposit of Petit Anse,* pp. 17-18; Taylor, *Destruction and Reconstruction,* p. 114; and Lucas, "Rock Salt in Louisiana," *Transactions of the American Institute of Mining Engineers,* volume XXIX, pp. 464-65.

43. Its thickness is even yet not known, as drills have penetrated it to a depth of 2,200 feet without passing through it. Based on the personal statement of the superintendent at the mine to the writer.

44. Hilgard, *The Salines of Louisiana,* published in the *Mineral Resources of the United States, Geological Survey,* p. 559.

45. Chemical analysis revealed a composition almost entirely pure:

| | |
|---|---|
| sodium chloride | 98.8823 |
| sulphate of lime | .7825 |
| magnesium chloride | .0030 |
| calcium chloride | .0036 |
| moisture | .03286 |

Analysis of Gössmann, *On the Rock Salt Deposit of Petit Anse,* p. 17.

The above analysis agrees in the essential of the proportion of sodium chloride with the analysis made by Hilgard, the Mississippi state geologist, *The Daily Mississippian* (Jackson, Miss.), October 10, 1862. Commercial salt is usually contaminated to the extent of 2 to 5% or even more with foreign ingredients, whereas this rock salt deposit has only the trifling amount of one-tenth per cent of gypsum.

46. Latham, *op. cit.,* p. 185.

47. See the report of an agent sent from Mississippi, *O.W.R.,* series I, volume LII, part 2, pp. 355-56.

48. The discovery of a salt mine was reported at Opelika also. See *The Charleston Daily Courier,* August 12, 1862.

49. See below chapter VI.

50. *O.W.R.,* series I, volume LII, part 2, p. 354.

51. See *The New York Times,* April 27, 1863, for modes of transportation.

52. No attempt is made to burden the reader with every little salt work or well which has come to the attention of the writer. It must be recognized, however, that there were doubtless numerous places, besides those mentioned, where individuals were making small supplies of which no record has been preserved.

See map for the various places throughout the confederacy where salt was made.

## CHAPTER III

1. Entry of British salt into the north continued, naturally, during the war. Of the 937,833 tons produced from the fields of Cheshire, Worcestershire, and Belfast in 1864, 86,208 tons were exported to America. Latham, *Black and White,* p. 189; "Statistics of Trade and Commerce", *The Merchants Magazine and Commercial Review,* volume 46, p. 545; *On the Rock Salt Deposit of Petit Anse,* p. 25, (note).

2. G. W. Brame to Governor Pettus, November 28, 1861; Pettus Papers, volume 54, Mississippi Archives.

A very humble woman, who had two sons in the service and four small children to support, turned in her despair to the chief executive. Her English, though hardly impeccable, is wholly intelligible: "What hogs we have to make our meat, we can't get salt to salt it." Mrs. Sarah Brown to Governor Pettus, December 18, 1861. *Ibid.*

3. G. W. Brame to Governor Pettus, November 28, 1861. He had a few days earlier, November 14, 1861, written Governor Moore in the same strain. A. B. Moore Letter File, Alabama Archives, Petitions on this subject appeared, as would be expected, in practically every legislature through the south. Note *Virginia House Journal,* Called Session, November, 1861, p. 124.

4. *Daily Vicksburg Whig,* November 15, 1861. It added to the citizen's sense of grievance that it was currently, if possibly erroneously, reported that salt stocks in New Orleans were sufficient to supply the confederacy for from one to two years.

5. Charles Manly to Governor Vance, November 13, 1861, and June 24, 1862, Swain Manuscripts, 1770-1869, North Carolina Historical Commission.

6. Published in the *South Western Baptist* (Tuskegee, Alabama), October 17, 1861. See also Governor Letcher's burning words to the Virginia assembly in December, 1863, *Virginia Senate Journal,* 1863-64, pp. 11-12.

7. See the account given in the governor's annual message, 1862, *Florida Senate Journal,* December, 1862, p. 49. For the orders and correspondence in regard to this seizure, see Documents accompanying the governor's message, *Ibid.*, pp. 72-76.

8. *Florida Senate Journal,* Session of 1863, pp. 20-21.

9. *The Confederate Records of the State of Georgia,* volume II, p. 215. The date is April 21, 1862.

10. *Session Laws of Georgia,* Extra Session, March-April, 1863, No. 106. Governor Brown stated in 1862 that while there were no written contracts with the salt companies that their production was not to be sold on speculation, that fact was well understood between himself and them. *Georgia House Journal,* 1862, p. 64.

11. Letter Books-Governors', 1861-65, p. 361, Georgia Archives.

12. A group of Mississippians, for instance, who had sent an agent to the Alabama salt wells to buy salt, feared with justice that the sellers would take advantage of an interdict laid by the governor of Alabama on the removal of salt from the state, issued subsequent to the sale and payment for the salt, to repudiate their contract. See letter of A. B. Longstreet to Governor Pettus, (no date), Pettus Papers, volume 58, Mississippi Archives.

13. Upon Governor Pettus's complaint to that effect to the president of the Southern Railroad Company, the latter declared that he had issued imperative orders to give preference to salt intended for the use of families and neighborhoods over that to be sold, especially that belonging to speculators. Pettus Papers, volume 58.

14. *Virginia House Journal,* Session of 1863-64, p. 8.

15. *Virginia Senate Journal,* Extra Session, 1863, Appendix, Document

No. 10, p. 20. Report of the Joint Committee on Salt in Respect to salt for Georgia.

A North Carolinian wrote on June 24, 1862, to the governor of that state that a speculator had slipped over to Raleigh and had bought up every pound of salt and bacon. Charles Manly to Governor Vance, Swain Manuscripts, North Carolina Historical Commission.

16. November 22, 1861. The following, printed in the issue of November 18, 1861, of the above paper is equally suggestive: "Napoleon Bonaparte found himself confronted at one period by precisely this state of things. His remedy for the evil was characteristic of himself; but it was very effective. He seized the produce and he shot the speculators. They got no pay at all for their property, except a free passage out of a world which they had outraged. Some effective measures of the sort will have to be taken with the speculators in salt, and in other articles of human food and venture.—".

17. *Ibid.*, November 25, 1861; *The Daily Press* (Nashville), June 10, 1863. Individuals did not hesitate to name offenders to the authorities. Such a list of firms was sent to Governor Vance, including the "Worth's," meaning apparently the men who in turn became salt commissioner. Executive Papers, Z. B. Vance, North Carolina Historical Commission. See also *Richmond Enquirer,* November 8, 1862; *The Western Democrat* (Charlotte, North Carolina), December 17, 1861; Mrs. L. P. Shryock to Governor Vance, April 7, 1863, Executive Papers, Z. B. Vance, North Carolina Historical Commission.

18. The entry of some salt into Houston, Texas, was noted in *The Clarke County Democrat,* (Weekly, Grove Hill, Ala.), December 5, 1861.

Note the advertisement in the *Southern Confederacy* (Atlanta) of March 30, 1862:

Salt! Salt!
80 sacks Liverpool and Virginia Salt
For sale at ——

19. *The Correspondence of Jonathan Worth,* volume I, p. 165. Many letters of complaint are found in the Pettus Papers, volumes 58, 59, and 61, Mississippi Archives; and in the Executive Correspondence, Alabama Archives.

20. David Rosenberg to Governor Pettus, December 15, 1862, Pettus Papers, series E, volume 58, Mississippi Archives.

21. "Report Relative to the Release of Salt held by the Confederate Government," *Virginia Documents,* 1861-62, Document No. 39, p. 8.

It might be noted in passing that salt was not the only article which was scarce. One letter encountered in the Pettus Papers is written on a torn, stained tax receipt, on which the writer had added superfluously, "Paper is scarce." The writer also found in the same collection a queer, gray-brown paper which had been used in the office of the Board of Police of Pontitoc County in September, 1864. So poor was the paper and so poor the ink that it is almost impossible to decipher the record.

22. Executive Papers, Z. B. Vance, North Carolina Historical Commission.

23. "Report of the Joint Committee on Salt Relative to the Contract with Stuart, Buchanan and Company", *Virginia Documents,* 1863-64, Document No. 26, p. 4.

24. One county agent, J. L. Ward, stated, March 8, 1865, that his predecessor had failed totally to supply the demands of the people, and wanted to know if there were any chance to get salt shipped from Saltville or from any other place. Governor Vance wrote on the paper as direction for the reply: "No chance to get it from Va. now." Z. B. Vance Papers, volume VII, North Carolina Historical Commission.

25. *O.W.R.*, series I, volume LI, part 2, p. 312; volume XII, part 3, p. 890; volume X, part 2, p. 574; volume XIX, part 2, p. 122; volume X, part 2, p. 568; volume XX, part 2, p. 156; volume XVII, part 2, pp. 802, 817; volume XXIV, part 3, p. 596; volume XLI, part 2, p. 458; volume XXII, part 2, p. 785.

26. *Ibid.*, volume XXIV, part I, p. 289.

27. A North Carolina citizen also wrote to Governor Vance that many were compelled to sell their hogs for lack of salt. A. McDowall to Governor Vance, March 7, 1863, Executive Papers, Z. B. Vance, North Carolina Historical Commission.

28. *O.W.R.*, series I, volume XII, part I, p. 733. See also volume XLI, part 2, p. 458.

29. *Ibid.*, volume XIII, pp. 445, 461; volume XVI, part 2, p. 117; volume XVII, part I, p. 499; volume IX, p. 338. People lacked salt even for the corn bread. *Ibid.*, volume XVII, part 1, p. 499.

30. *The Confederate Records of the State of Georgia,* volume II, p. 231. Governor Brown's opinion was that only one-half of the meat of the state could be saved for the season 1862-63. *Ibid.*, pp. 237-38.

31. *O.W.R.*, series I, volume LII, part 2, p. 354.

32. *Ibid.*, pp. 383-84.

33. *The American Annual Cyclopedia,* 1862, p. 800.

34. *Ibid.*, p. 661.

35. *O.W.R.*, series I, volume XVII, part 2, p. 261.

36. Reported to the confederate congress, *Ibid.*, series IV, volume II, pp. 854-55.

37. Avary, *A Virginia Girl in the Civil War,* p. 350.

38. Clay-Clopton, *A Belle of the Fifties,* pp. 223-24. She was writing of 1864.

39. *Our Women in the War,* No. 24, p. 141; Pickett, *The Heart of a Soldier,* p. 12; Townsend, *Campaign of a Non-Combatant,* p. 40.

40. N. A. Boylan to Governor Vance, July 4, 1863, Executive Papers, Z. B. Vance, North Carolina Historical Commission. See also T. G. Davidson to Governor Pettus, December 20, 1862, Pettus Papers, volume 58, Mississippi Archives.

41. *The Correspondence of Jonathan Worth,* volume I, p. 357; R. T. Scott to J. A. Graham, Auditor, May 7, 1863, Executive Correspondence, series W. Alabama Archives.

42. *O.W.R.*, series I, volume XVIII, pp. 853, 897-98; Governor Vance to Secretary Seddon, February 28, 1863, Vance Letter Book, January-April, 1863, North Carolina Historical Commission; Ashe, *History of North Carolina,* volume II, p. 926.

43. *On the Rock Salt Deposit of Petit Anse,* p. 26. In this instance the bushel was computed at seventy pounds.

44. *The Clarke County Democrat* (Weekly, Grove Hill, Ala.), Aug. 29, October 3, 1861. That paper quoted the rumor that powerful combina-

tions of capitalists wished to prevent the transportation of salt. "Thus while Northern Yankees blockade our sea-ports," raged that editor, "home Yankees blockade our interior channels of trade. We are fighting the Northern Yankees with canon and bayonet. What sort of weapons should we use against the home Yankee, their aides and abettors?" *The Richmond Dispatch,* October 1, 1861.

45. *The Savannah Daily Republican,* January 8, 1862; Jones, *A Rebel War Clerk's Diary,* volume I, pp. 183, 194. Salt was commanding about the same price in North Carolina at that time, but sold higher in Memphis in June at the soaring price of one hundred dollars a sack or sixty-six cents a pound. McMaster, *History of the People of the United States during Lincoln's Administration,* p. 329.

46. *The Confederate Records of the State of Georgia,* volume II, p. 280.

47. Fleming, *Civil War and Reconstruction in Alabama,* p. 158. The low price cited in the text was noted at Arkadelphia and Lake Loudon, Louisiana. *O.W.R.,* series I, volume XIII, p. 771.

48. *The Confederate Records of the State of Georgia,* volume II, p. 429; *O.W.R.,* series IV, volume II, pp. 181-82; *Session Laws of Virginia,* Called Session, September-October, 1863, Joint Resolution, No. 6; Davis, *The Civil War and Reconstruction in Florida,* p. 207.

49. Everything else was, of course, in proportion. See Varina Davis, *Jefferson Davis, A Memoir,* volume II, pp. 526-35; Moore, *History of North Carolina,* volume II, p. 249.

50. *O.W.R.,* series I, volume XXXII, part 2, p. 658; *A Calendar of Confederate Papers,* p. 391; *The Confederate Records of the State of Georgia,* volume II, pp. 692-95. Comparison should be made with the rate of 1862-63. *Ibid.,* volume II, pp. 237-38. See also Pettus Papers, series G, 201, Mississippi Archives.

51. Miller, *History of Alabama,* p. 230. Price in terms of other commodities is discussed later.

52. Governor Brown to President Davis, March 28, 1865. Letter Books-Governors', 1861-65, p. 751, Georgia Archives; Eggleston, *A Rebel's Recollections,* p. 92.

53. Dodge, "Domestic Economy in the Confederacy", *Atlantic Monthly,* volume 58, p. 231. Liverpool salt sold at $25 in Richmond, at $60 in Beaufort, North Carolina, *O.W.R.,* series IV, volume I, p. 1010. See *The New York Times,* April 18, 1864.

54. *O.W.R.,* series I, volume XV, p. 782.

55. *Ibid.,* volume XIII, p. 885.

56. *Ibid.,* volume XXVI, part 2, p. 206.

57. *Session Laws of Mississippi,* November-December Session, 1861, chapter 134; Garner, "The State Government of Mississippi during the Civil War", *Political Science Quarterly,* volume XVI, p. 294.

58. Letter Books—Governors', p. 163, Georgia Archives. The owner was allowed to sell salt, wheat, flour, bacon, lard, cotton, osnaburgs, leather, and shoes at an advance of sixty per cent over the price of April, 1861, but all speculative prices were forbidden, *The American Annual Cyclopedia,* 1861, p. 342. Osnaburg was a very coarse cotton cloth.

59. *Ibid.,* 1862, p. 553; *Session Laws of South Carolina,* February-April, 1863, No. 4621.

60. It is interesting to note in this connection that in earlier civilizations salt was used alongside of metal money. In the Roman army, for instance, allowances of salt were made to officers and men, from which their *solarium* was converted into an allowance of money for salt. Hence, was derived the word for salary. The *United States Bureau of Mines*, Bulletin No. 146 has an excellent presentation of the entire subject of salt production in the United States.

61. F. Vaughan to Governor Pettus, October 19, 1862, Pettus Papers, volume 58, Mississippi Archives; Wilson Ashley to John Long, November 8, 1862, series W, Alabama Archives. Toward the close of the war flour in small quantities was still being brought down to Florida from Georgia to be exchanged for salt, which was up to the end of the war being made in sheltered bays. Brevard, *A History of Florida,* volume II, p. 115.

Other instances of barter are to be found in letters from Turner to Governor Pettus, July 7, 1863, Pettus Papers, volume 61; Governor Clark to Governor Letcher, May 5, 1862, Letter Book, 1861-62, Governor Clark, p. 311, North Carolina Historical Commission; contract of Elisha Bonner with Stuart, Buchanan Company, October 7, 1862, Executive Papers, Z. B. Vance; W. M. Hicks to Governor Vance, April 25, 1863, *Ibid.*

62. See telegram of Superintendent Long of the Alabama State Salt Works to Governor Shorter, September 25, 1862, and Shorter's reply, series W, Alabama Archives; *Session Laws of Mississippi,* Regular Session, 1863, chapter XLV, section 8; Abstract C of Woolsey's Report for the second quarter of 1865, Alabama Archives.

63. Davis, *Civil War and Reconstruction in Florida,* p. 204; *Florida Senate Journal.* 1863, pp. 20-21; *O.W.R.,* series I, volume XIII, pp. 885-86.

64. Report of N. W. Woodfin, salt agent, to Governor Vance, November 27, 1862, *Legislative Documents of North Carolina,* 1862-63. No. 13, p. 3. Woodfin to Ex-Governor Swain, November 17, 1862, Walter Clark Manuscripts, volume IV, North Carolina Historical Commission.

65. Northrop, commissary-general, reported to Breckenridge, Secretary of War, on February 9, 1865, as follows on this subject: "The contract made with Messrs. Stuart, Buchanan, and Company, was for 45,000 bushels per month in excess of the estimated demand for the army. This was made in view of the foreseen deficiency of money, to obtain meat by supplying salt for barter. After the contract was made and approved by the honorable Secretary of War, he, in spite of my remonstrances, and, in my judgment, contrary to the interest of the Government, annulled it. Believing this to be unjust and prejudicial to the interest of the Government, I refused to take any action in the matter, and he then assumed the making of a contract with the state of Virginia which I had considered and declined to make." *O.W.R.,* series I, volume XLVI, part 2, pp. 1221-22.

66. See below, Chapter IX, pp. 158-59.

67. Von Borcke, *Zwei Jahre im Sattel und am Feinde,* volume II, p. 105.

68. "Memoranda Concerning the Geological Survey", *Mississippi Senate Journal,* Called Session, 1862, Appendix, pp. 90, 94.

Hilgard tested salt and brine from Rankin, Panola, Hinds, Pike, Scott, Marshall, and Holmes Counties. *Ibid.,* pp. 92-93. It would not appear that he tested the Virginia salt or any of the foreign salts, probably as there was little hope of obtaining them for Mississippi. Veatch found the brine from King's well in Louisiana, while slightly inferior in sodium chloride to that from Lake Bisteneau, superior to Kanawha salt—seven per cent impurities as against twenty to thirty per cent. "The Salines of North Louisiana," p. 93.

69. Governor Clark transmitted to the legislature Professor Emmons's communication, which no doubt contained interesting information, but, unfortunately, the writer has been unable to find it among the North Carolina manuscripts and she has been unable to locate any other paper which would reveal what he had in mind. The allusion occurs in Letter Book, 1861-62, Governor Clark, p. 211, North Carolina Historical Commission.

70. *The Clarke County Democrat,* (Weekly, Grove Hill, Alabama), November 13, 1862. The same thing was printed in *The South Carolinian* of Columbia, South Carolina, November 10, 1862.

71. Dodge, "Domestic Economy in the Confederacy", *Atlantic Monthly,* volume 58, p. 231.

72. *O.W.R.,* series IV, volume II, p. 574. Of course, some was lost by inexperienced persons trying to cure meat by unfamiliar processes.

73. V. Davis, *Jefferson Davis, A Memoir,* volume II, p. 558, note.

74. "Report of the Joint Committee on Salt in Respect to Salt for Georgia", *Virginia Documents,* Extra Session, 1863, Document No. 10, pp. 17-18. The report is dated October 30, 1863, and is bound with the *Virginia Senate Journal and Documents* for that session.

The writer could not learn whether an apparent effort in North Carolina to utilize waste products of salt manufacture bore fruit or not. See *Private Laws of North Carolina,* Regular Session, 1864-65, chapter 18, section 2, which directs that the salt commissioner and salt agent of the state turn over all waste resulting from their salt activities to a company of newly-incorporated chemists.

75. Pickett, *The Heart of a Soldier,* p. 11; Hague, *A Blockaded Family,* pp. 38-39. It is hardly necessary to cite all the references where the writer found this story recorded, nor the many persons who gave it to her by word of mouth.

76. *The Charleston Daily Courier,* January 24, 1862; *The Charleston Mercury,* November 19, 1862; *South Carolina Women in the Confederacy,* volume I, p. 155.

77. Lunt, *A Woman's War-Time Journal,* p. 18. The army was not untouched by the zeal for economy. A circular in Tennessee early in 1863 ordered that all the salt received in mess-beef issues by the troops should be turned over to the artillery and quarter-masters of their respective commands for the use of horses. *O.W.R.,* series I, volume XXIII, part 2, p. 665.

78. "Minority Report of the Joint Committee on Salt Relative to Supplying the People with Salt", *Virginia Documents, December Session,* 1863, Document No. 28, p. 3. Found in *Message of the Governor and Accompanying Documents,* 1863. It is also printed as Document No. 6, bound with the *Virginia Senate Journal,* 1862-63, as an appendix to the governor's message. The reference occurs on p. 13.

79. *Hinds County Gazette* (Raymond, Mississippi), September 3, 1862.

The first rules were relatively prodigal of salt, as salt was to be rubbed on a fourth of an inch thick. *South Western Baptist* (Tuskegee, Alabama) December 5, 1861. See also the *Daily Vicksburg Whig,* January 11, 1862, and the really helpful article in the *South Western Baptist* of December 25, 1862, which brings out some of the scientific facts to be known by a person trying to preserve meat.

80. *Resources of the Southern Fields and Forests,* p. 372. Recipes for cooking salt beef in order to make it palatable appeared in the newspapers. See *Charleston Daily Courier,* December 28, 1861, where a recipe was vouched for by an old soldier with long experience in camp life.

An additional fact comes out in the records, showing the effect of weather conditions in different parts of the south on packing problems. An agent from Alabama pointed out that his state must put up her pork during the first suitable weather, as it often happened in that state that suitable weather came but once during the winter season, while the people of Virginia could kill their pork safely any week from the last of October till the last of March. This factor entered to embitter the quarrel between Virginia and her southern neighbors, as detailed in Chapter IX. *O.W.R.,* series I, volume LII, part 2, p. 385.

## CHAPTER IV

1. Their troubles, however, were far from over, for Governor Shorter forbade the removal of the salt from Alabama to other states, as set forth in Chapter VI. They thereupon appealed to their own governor, pleading the inviolability of contract. Pettus Papers, vol. 58, Mississippi Archives.

2. *The Daily Mississippian* (Jackson), November 11, 1862.

3. Dodge, "Domestic Economy in the Confederacy", *Atlantic Monthly,* vol. 58, p. 231.

4. One man was reported as contemplating establishing works on St. Andrew's Bay, Fla., large enough to supply one of the midland towns where he had been offered ninety dollars a barrel. Hopley, *Life in the South,* II, p. 308.

5. *The Charleston Mercury,* June 27, 1862.

6. *The Charleston Daily Courier,* January 28, 1862. An auction sale of the samples was held. Displaying samples of various products in newspaper windows seems to have been a practice of the times. See also *Daily Richmond Examiner,* November 26, 1861.

7. *The Daily Mississippian* (Jackson), June 27, 1862.

8. One man reported concerning the Santrell Salt Works, evidently near Mobile, on April 4, 1864, "You wanted to know how we are getting along, only tolerable. We have sold 50 sacks at $45 and we have about 30 on hand. We can't get barrels or sacks here. . . . I have sent to Mobile twice and failed." Candy to J. B. Long, April 24, 1864, series W., Alabama Archives.

A Mississippian, after securing a lease in Alabama, complained of the lack of boilers. John H. Evans to Governor Pettus, (undated). Pettus Papers, vol. 61.

9. See an advertisement in *The Daily Mississippian* (Jackson), July 19, 1862.

Note also the following, headed Kidd's Salt Works, Clarke County, Alabama:

"Parties who wish to make Salt, will find it to their interest to go to those remarkable Salt Springs and Lakes. The undersigned can be seen for a few hours on Monday morning, 21st inst., at J. W. Robinson and Co. He will lease the use of salt water, and give certificates therefor. Parties can in this way secure the right to use the water, as they will be guaranteed the right to locate when presented.

"This locality is within 12 hours run of the city of Mobile, is ¾ of a mile of the Bigbee River," *The Daily Mississippian,* July 19, 1862.

10. Pettus Papers, volume 61, Mississippi Archives.

11. Bostick and Co. to Governor Shorter, September 28, 1863, series W, Alabama Archives. For a case perhaps more disinterested see James W. Osborne to Governor Vance, April 6, 1863, Z. B. Vance Papers, volume II, North Carolina Historical Commission. See also W. C. Askow and A. Jaques to Governor Vance, February 12, 1863, Executive Papers, Z. B. Vance; W. D. Smith to Governor Vance, February 9, 1863, *Ibid.;* Thos. Miller to Governor Vance, February 4, 1863, *Ibid.*; also the offer of three Virginians to the Joint Committee on Salt, dated September 22, 1862; and One Package of Papers Relating to Salt, 1860-65, (Mss.) Virginia Archives; and Crafton Baker to Governor Pettus, June 21, 1863, Pettus Papers, volume 61, Mississippi Archives.

12. J. M. Wilson to Governor Vance, March 31, 1863, Executive Papers, Z. B. Vance, North Carolina Historical Commission. The property was no longer in the possession of his family.

13. *The Correspondence of Jonathan Worth,* volume I, p. 166-67. See also *The Weekly Raleigh Register,* December 18, 1861, concerning a salt "fountain" near Charlotte which flowed for years.

14. Letter of W. C. Kerr to Governor Vance of June 2, 1863, Executive Papers, Z. B. Vance, North Carolina Historical Commission.

15. He indicated another way to measure the salinity of water by an hydrometer, and offered, to prove his point, the following comparative figures based on hydrometer readings:

| | | | | | | | |
|---|---|---|---|---|---|---|---|
| sea water | 21-2 | 350-400 | gallons | yield | 1 | bushel | salt |
| Saltville brine | 20-25 | 30 | " | " | 1 | " | " |
| South Carolina beach water | 10-18 | 60-100 | " | " | 1 | " | " |

W. C. Kerr to Governor Vance, November 28, 1864, Executive Papers, Z. B. Vance, North Carolina Historical Commission.

Professor Kerr's company had already appealed to Governor Vance for a detail of conscripts in April, 1863, as it was impossible to hire negroes. James N. Osborne to Governor Vance, April 6, 1863, Z. B. Vance Papers, volume II,

16. Rich Archer to Governor Pettus, December 1, 1862, Pettus Papers, volume 58, Mississippi Archives. The writer of the letter proposed to furnish the labor, the state to supply a scientist to direct the search and to pay all the other costs, but the owner wished the right to work the mine at a reasonable rental.

17. Vardin to Governor Pettus, December 30, 1862. *Ibid.* It is difficult to determine from the manuscript the spelling of this name.

18. W. R. Symons to Adj. General Wayne, April 30, 1862, a loose letter found in the Georgia Archives.

19. *Journal of the Senate of the Confederate States of America,* 1 Congress, 1 Session, April 18-October 13, 1862, published as United States *Senate Documents,* 56 Congress, 2 Session, No. 234. The bill was introduced August 26, 1862, referred August 28, and a substitute reported September 3. See *Ibid.,* volume II, pp. 238, 248, 258. For the substitute see Unpublished Manuscripts in the War Department; "Proceedings of First Confederate Congress," *Southern Historical Society Papers,* volume XLV, pp. 19-20.

20. The rumor of a recently discovered salt mine was alluded to in the debate where it was said that "pure salt existed in boundless quantity". *Ibid.,* pp. 240-41. See also *House Journal of the Confederate States,* volume V, pp. 317-18.

21. *Ibid.,* volume VI, pp. 546, 829.

22. "Report of the Committee on Salt Supply," pp. 1-4. This report exists in the Virginia Archives in printed form. One member of the committee thought that the salt could be used in its impure state for stock, and for curing bacon by merely being pulverized. *Ibid.,* p. 7.

23. Opinion of A. Findlay, transmitted to the Committee on Salt, *Ibid.,* p. 7.

24. Schawb, *The Confederate States of America,* pp. 242, 291; *The Statutes at Large of the Confederate States of America,* Provisional Congress, 2 Session, chapter 44, Schedule F; *Ibid.,* First Congress, (Regular), 3 Session, chapter XXXVIII, section 1.

25. *Ibid.,* 1 Congress, 4 Session, chapter 66, section 9.

26. *Ibid.,* 1 Congress, 1 Session, chapter XXXI.

27. At this early period only employees on railroads and river routes of transportation, employees in mines, furnaces, and foundries were named because of their industrial importance. Superintendents and operatives in wool and cotton factories could be exempted at the discretion of the secretary of war. *O.W.R.,* series IV, volume I, p. 1081; *The Statutes at Large of the Confederate States of America,* 1 Congress, 1 Session, chapter LXXIV. It did not even embrace workers in the one salt mine of the confederacy, as that discovery still lacked a few days of being made.

The failure to include salt workers generally greatly perturbed some people. See *The Correspondence of Jonathan Worth,* volume I, p. 163.

28. *O.W.R.,* series IV, volume I, p. 1127. See General Orders No. 50, *Ibid.,* volume II, p. 8. They also appeared in the Richmond papers.

29. See the correspondence between Gov. Shorter and Secretary Randolph of July 31, Aug. 4, Aug. 6, Aug. 7, and Sept. 10, 1862. One of Randolph's replies, that of Aug. 23, is missing from the Alabama files, but the purport is clear from Governor Shorter's letter of September 10, Executive Correspondence, Alabama Archives. In the course of the correspondence Governor Shorter went so far as to direct exemption of one of the salt makers from the draft. *Ibid.,* series W. Alabama Archives.

30. He condemned these cowards as "dodging from place to place to avoid being made conscripts, and say that they would rather die than to be disgraced by being made conscripts, and doubtless would as willingly

be hung as traitors as die in battle vindicating the rights of freemen." *O.W.R.*, series IV, volume II, p. 95.

31. *Ibid.*, pp. 45, 93-94; *Journal of the House of Representatives of the Confederate States of America,* volume V, pp. 343, 361, 379.

32. *The Statutes at Large of the Confederate States of America,* 1 Congress, 2 Session, chapter XLV; *O.W.R.*, series IV, volume II, pp. 161-162.

33. Gov. Shorter to G. K. Clopper, March 30, 1863, Executive Correspondence, Alabama Archives.

34. *O.W.R.*, series IV, volume II, p. 167.

35. Complaints were many of abuses in connection with exemptions. F. Thompson wrote Gov. Vance on Sept. 27, 1862, that he knew several men who were pretending to make saltpetre, but that the whole group of six men were making less than twelve pounds a month. Z. B. Vance Executive Papers, 1862, North Carolina Historical Commission.

36. *The Statutes at Large of the Confederate States of America,* I Congress, 4 Session, chapter LXV. It is also printed in *O.W.R.*, series IV, volume III, pp. 178-181.

37. *The Statutes at Large of the Confederate States,* I Congress, 4 Session, chapter LXV, section XI.

38. *O.W.R.*, series IV, volume III, p. 358.

39. For a full study of the statutory exemption question, see Moore, *Conscription and Conflict in the Confederacy,* chapters IV and V.

40. *O.W.R.*, series IV, volume II, pp. 161, 553-54; *The Statutes at Large of the Confederate States of America,* I Congress, 3 Session, chapter LXXX, section 4; *Ibid.*, I Congress, 4 Session, chapter LXV, sections 10, 11.

41. *Alabama House Journal,* Regular Session, 1864, pp. 371-73.

42. In this particular case he had first written rather petulantly that he did not see what power he had over the matter. "I have no power to compel the Secretary of War to grant the exemptions." However, before long he was found addressing President Davis in their behalf. See Gov. Brown's reply to Seago, Palmer Co., Jan. 31, 1864, Governors' Letter Books, 1861-1865, pp. 591, 598, Georgia Archives.

43. He provided a man engaged in furnishing wood to an engine with a certificate. *Ibid.*, p. 676.

44. *Ibid.*, Governors' Letter Books, 1861-1865, pp. 607, 716, 717, 718; *The Confederate Records of the State of Georgia,* volume II, p. 695.

45. *O.W.R.*, series IV, volume III, pp. 851, 867, 873, 1102.

Other instances of appeals for exemption are found in the files of the North Carolina Historical Commission; Worth to Vance, Sept. 6, 1864, where the former asks exemption for some barrel-makers, Z. B. Vance Executive Papers, September, 1864; McNeill to Gov. Vance, March 17, 1864, where exemption is sought for a county salt agent, *Ibid.*, March, 1864; and H. C. Lee to Gov. Vance, April 30, 1863, where exemption is asked in order to haul salt, *Ibid.*

At one time Georgia had 15,000 men on the exemption lists, but all except 1,400 were in the militia, Governors' Letter Books, 1861-1865, p. 738, Georgia Archives.

46. The figures as furnished by Preston, Superintendent of Conscripts, are interesting as affording comparison between the several states:

| | | | |
|---|---|---|---|
| Virginia | 41 | Mississippi | 0 |
| North Carolina | 5 | Florida | 15 |
| South Carolina | 12 | East Tennessee | 7 |
| Georgia | 29 | East Louisiana | 0 |
| Alabama | 4 | | |

*O.W.R.*, series IV, volume III, pp. 1104-5.

47. Pumphry, *Memories,* p. 137.

48. Act of Oct. 11, 1862. *The Statutes at Large of the Confederate States of America,* I Congress, 2 Session, chapter XLV, section I.

49. The first yearly meeting of the Friends after the passage of the act was held at New Garden, Guilford County, in 1862. A minute of the meeting declares that while the Friends would pay all taxes imposed on all citizens, they could not "conscientiously pay the specified tax, it being imposed upon us because of our principles, as the price exacted for religious liberty." Many members did pay the exemption tax and the meeting was not disposed to censure them. Cartland, *Southern Heroes* or the *Friends in War Time,* pp. 140, 141.

It is an interesting fact that some North Carolina Quakers went to Indiana to escape conscription into the confederate army; but there they soon found themselves drafted into the union army, and so were forced to pay an exemption in the north. Weeks, *Southern Quakers and Slavery,* p. 306, note 1.

50. *North Carolina Ordinances of the State Convention,* 1861-2, Fourth Session, No. 34. Ratified May 12, 1862. See also *The Correspondence of Jonathan Worth,* I, p. 207, which reveals the perturbation of some Quakers at the possibility of a double tax.

It is entertaining that the central government felt the Quakers to be a sufficient menace to militaristic policies so that when a Joseph Newer, a Quaker from England, asked leave to enter the confederacy from the United States, Secretary Seddon denied to this missionary "free intercourse with our people." Seddon to Gov. Vance, Z. B. Vance Papers, VI, January 18, 1865, North Carolina Historical Commission.

51. Dated Sept. 19, 1862, Vance Letter Book, 1862-63, North Carolina Historical Commission.

52. *The Correspondence of Jonathan Worth,* I, pp. 354-55.

53. The salt captured from the United States could hardly be classified as one of the methods employed. Naturally, the confederate army seized such supplies whenever it could. See "Jackson's Valley Campaign", *Southern Historical Society Papers,* volume 43, p. 233, Note. Just as naturally, it sought to save its own stores of salt from falling into the hands of the foe. See *O.W.R.,* series I, volume XIX, part 2, 612; volume XXXII, part 2, p. 664.

54. *Journal of the Senate of the Commonwealth of Virginia,* Reg. Sess., Dec., 1861. Appendix; "Report of the Joint Committee Appointed to Visit the Confederate Authorities in Regard to Supply of Salt," Document 7, p. 7. (Statement of Northrop, Commissary General.)

55. One Package of Papers Relating to Salt, Virginia Archives, 1860-65; "Minority Report of the Joint Committee Appointed to Correspond with the Lessees of the Salt Works", *Virginia Documents.* Document No. 47, 1861-62, p. 3.

A contract apparently entered into by the central government with the Virginia proprietors on February 5, 1863, for the delivery of 300,000 bushels by April, 1864, seems slightly contradictory with the above, which would call for only 264,000 bushels. Under the contract last alluded to the government was to advance $100,000 for the salt, guaranteed by the company's bond. This advance of money brought some criticism, as the president interpreted the act of congress, authorizing advances for the purchase of munitions, to include the commissary department. The contract is found among the unpublished manuscripts of the War Department.

56. *Georgia Journal and Documents of the Senate,* Extra Sess., 1862, Document No. 10, p. 27.

57. Governors' Letter Books, 1861-65, p. 114, Georgia Archives.

58. Veatch, *The Salines of North Louisiana,* Part VI, p. 59.

59. *Ibid.,* p. 84.

60. It has been a matter of disappointment to the writer that it has proved possible to locate only a very few of the government contracts.

61. For one such instance see a union report to General Thomas from Kentucky, December 18, 1861, *O.W.R.,* series I, volume VII, p. 505.

62. D. C. Green, Quarter Master General, to Gov. Shorter, May 27, 1862, and Shorter's reply; Major F. W. Frances to Gov. Shorter, July 29, 1862, Executive Correspondence, Alabama Archives.

63. See letter of Gov. Shorter to Gen. J. G. L. Henry and to F. L. Johnson and Co., both of Sept. 8, 1862, Executive Correspondence, Alabama Archives. The governor was concerned for fear citizens might think that the state was speculating upon their necessities.

64. Governors' Letter Books, 1861-65, p. 565, Georgia Archives.

65. *O.W.R.,* series I, volume VII, pp. 60-61, p. 756.

66. Mr. Chew, the Mississippi agent, sent to New Iberia to secure a supply of salt for that state, reported on Sept. 16, 1862, a gloomy outlook if the country were to depend on the activities of the owner of the mine. But one shaft had been sunk, and scarcely one-third of the work completed on another. He had only about sixty hands at work in and about the mine. The only hope, in this agent's opinion, lay in seizure by General Taylor for the central government. *Ibid.,* volume LII, part 2, pp. 355-56. This report was sent by Governor Pettus to President Davis, and transmitted by the latter through the secretary of war to the commissary general. See Pettus to Pres. Davis, Sept. 20, *Ibid.,* p. 354; Davis's reply of September 25, *Ibid.,* volume XVII, part 2, p. 713; and of October 25, *Ibid.,* volume LII, part 2, pp. 382-83.

67. *Ibid.*

68. Several states seem also to have sunk pits of their own; Mississippi, Alabama, and Georgia each owned one, while the county of Opelousas opened one. *Rock Salt Deposit of Petit Anse,* p. 31.

69. This account is based on Latham, *Black and White,* pp. 187-88 and on Taylor, *Destruction and Reconstruction,* p. 114.

70. The story of the several raids and of the final capture of the mine will be told in chapter XII.

71. Davis, *op. cit.,* pp. 207, 208 (note); *The American Annual Cyclopedia,* 1864, p. 378.

See below Chapter XI. The earliest hint that the Richmond government was engaged in the business of salt manufacture is found in the *Official Naval Records* in 1862, volume 19, pp. 410-411.

72. Veatch, *op. cit.,* pp. 84, 67.

73. Special Orders, No. 43, issued from Little Rock, Ark., Oct. 1, 1862, *O.W.R.,* series I, volume XIII, p. 885. While it was suggested to Sec. Randolph that the government take over some of the salines of North Louisiana, there is no reason to think it was done. *Ibid.,* volume XV, p. 779. It is interesting to find this precedent for the expropriation pursued by the United States in the World War in regard to the railroads.

These instances of salt production by direct government action differ from the earlier case of action by military authority related on p. 70 as there the wells were acquired by contract and purchase, here by preëmption.

74. See a letter from several citizens to Gov. Flanagin of Arkansas under date of July 25, 1863, *Ibid.,* volume XXII, part 2, p. 947.

75. *Ibid.,* volume XXVI, part 2, p. 354; volume LIII, p. 966; volume XLI, part 4, p. 33; volume VII, p. 756; and volume XXII, part 2, p. 190.

76. See above pp. 15-16.

77. He had so impressed the editor of *De Bow's Review* that several articles on the subject of the manufacture of salt by solar evaporation appeared in that magazine. See "New Salt Manufacture of the Confederate States", New Series, volume 31, pp. 442-46. Already in 1855 an article had appeared on the subject. volume XVIII, New Series, p. 538.

78. The method was, of course, not unknown in America, for at Syracuse a certain method of solar evaporation was in use, which, however, produced a very coarse grain.

79. His arguments are summarized in *De Bow's Review,* volume 26, pp. 119-20. (1859.) It is of passing interest that in 1776 an ordinance in Virginia provided for salt manufacture in several of the eastern counties of that state by a process of solar evaporation as well as by boiling from sea-water. *Senate Journal of Virginia,* Extra Sess., April, 1862, p. 81.

80. *Daily Richmond Examiner,* Nov. 13, 1861. The issue of Nov. 22 declares that the works had been established and would be turning out large quantities of salt by Christmas of that year. The prospectus is given in full in the article in *De Bow's Review,* volume 31, pp. 445-46.

81. *Alabama Session Laws,* Second Called Sess., 1861, No. 147, sec. 9. Ratified by the governor Dec. 3, 1861.

82. *Session Laws of Louisiana,* 1861-62, No. 115. Approved Jan. 23, 1862.

83. The writer is indebted for the above fascinating story to Veatch, "The Salines of North Louisiana." Part VI, pp. 84-85.

84. "Supplement a la géologie pratique de la Louisiane", *Bulletin de la Société Géologique de France,* Tome 20, 2e série, pp. 542-43.

85. This fact is established from the correspondence of Gov. Shorter with Col. Northrop, commissary-general, of Dec. 25, 1862. "When Prof. Thomassy revisits our salt lands, and makes the requisite survey, I will protect him in the uninterrupted enjoyment of such location as we may agree upon, and the operations shall proceed for the Confederate Govt. as contemplated by you." Executive Correspondence, Alabama Archives.

86. *Session Laws of Alabama,* Second Regular Annual Session, 1862, No. 37.

87. Telegram of December 16, 1862, series W, Alabama Archives.

88. Letter of Governor Shorter to Secretary Seddon, December 25, 1862. This letter is a reply to one from Seddon to the governor of December 17, which, unfortunately, is missing from the Alabama files. Executive Correspondence, Alabama Archives. See also his letter of the same date to Colonel L. B. Northrop, Commissary-General, *Ibid.*

89. If the government had undertaken the project, some mention of it would certainly have been made in the numerous works and records which discuss the salt activities in Alabama during the war. See below chapter VI. T. L. Head states that Alabama seriously considered trying the experiment of making salt by natural evaporation, and that at the close of the war excavation for a vat for this purpose was in process of construction at the Upper Works, Sketch of Historic Clarke County, a manuscript work never published. Mr. Head informs the writer personally that he had been given this information by Harrison Barnes, an old negro who had worked at the salt wells.

90. *O.W.R.*, series I, volume XX, part 2, pp. 486-87.

## CHAPTER V

1. *Mississippi Senate Journal,* Called Sess., Dec., 1862, pp. 8, 9.

2. It was urged that evidences of the presence of salt cropped out in many places in Georgia. It was thought to be associated with iron. *Southern Confederacy* (Atlanta), April 6, 1862.

3. E. W. Hilgard, *Memoranda Concerning the Geological Survey,* Appendix to the *Mississippi Senate Journal,* Called Sess., 1862, pp. 89-90.

4. *Ibid.,* p. 90. See above.

5. See Gov. Clark's letter to Milton Worth of July 21, 1862, "Let me therefore urge you to use the authority and means in your hands to bore in that region and do so immediately * * * *. If you can come up immediately, I want to leave no expedient untried for the supply of salt and I will render you every assistance in my power." Letter Book, 1861-62, Gov. Clark, pp. 378-79, North Carolina Historical Commission.

6. *Public Laws of the State of North Carolina,* Reg. Sess., 1864-65, Joint Resolution, p. 73. Ratified Dec. 23, 1864.

7. Package of Papers Relating to Salt, 1860-65 (MSS.), Virginia Archives. The idea did not seem unreasonable, as Botetourt County is not far from Saltville or from the Mercer Wells in West Virginia.

8. *Florida Session Laws,* Regular Sess., 1861, No. 1. Interestingly enough, this ordinance was repealed by the Constitutional Convention which reconvened in Jan., 1862, No. 56.

9. *Ibid.,* Sec. 4. Section 5 required the suspension of the act as soon as in the judgment of the governor the exigency for its continuance had passed.

The next year an act provided a penalty for purchasing under false representation as a state agent, *Florida Session Laws.* First Session, Twelfth Assembly, 1862, No. 44.

10. The case is an odd one, for the merchant declined to accept the money from the sale, and, after agreeing to receive drafts for the amount, left the sum in the treasury without presenting the drafts. Prob-

ably he was hoping to secure a settlement on higher terms through a court. *Florida Senate Journal,* 2 Sess., 1862, pp. 49-50.

11. *The Confederate Records of the State of Georgia,* II, p. 145.

12. Governor's Letter Books, 1861-65, p. 137, Georgia Archives.

The governor argued with justice, "if the constituted authorities do not interfere, but will pay, on the part of the State, the high prices, demanded by unpatriotic speculators, the cost of the supplies necessary to maintain our army will soon swell the public debt to an enormous burden, and as the high prices paid by the State will control the markets and compel her citizens to pay as much, provisions will be placed out of the reach of the poor, who labor for their daily bread, and much suffering and misery must result." *Ibid.,* p. 135. He had notified the commissary-general, Col. J. I. Whitaker, of his intended action on November 16. *Ibid.,* p. 132.

The assistant commissary seized 1,000 sacks of salt from A. K. Seago of Atlanta, paying the stipulated five dollars a sack. He then stored the lot with the former owner under the usual storage rates, but that individual later brought suit for the salt, which circumstance led to a conflict of civil and military authority. *Ibid.,* pp. 382-83; *The Confederate Records of the State of Georgia,* volume II, pp. 182-86.

13. See Gov. Brown's wire to the mayor on Nov. 21, Governors' Letter Books, 1861-65, p. 141; the mayor's reply of Nov. 25, p. 143; Brown's telegram of Nov. 27 with the mayor's response of Nov. 28, *Ibid.,* pp. 145, 114.

14. *Ibid.,* p. 163. The governor was obliged to modify his order slightly: any quantity of salt might be shipped to any point in the state for Georgia citizens. See also McMaster, *History of the People of the United States during Lincoln's Administration,* p. 329; and *The Savannah Daily Republican,* June 18, 1862.

15. Governors' Letter Book, 1861-65, p. 163, Georgia Archives.

16. *Ibid.,* p. 164.

17. *The Confederate Records of the State of Georgia,* II, p. 231. Meanwhile the governor, to protect executive action by law had recommended to the legislature which convened in November, 1861, authorization for the governor or a military officer under his command to seize any supplies necessary for the troops, wherever found, in return for a just compensation to be fixed by competent agents. Though not named, salt was, from the logic of preceding events, to be understood as included. *Georgia Senate Journal,* Reg. Sess., 1861, pp. 25-26. The records do not show that such a law was passed.

18. *Alabama Session Laws,* Regular Annual Session, 1861, No. 29.

19. Section 9 of the Extortion Act rose to vex Governor Shorter the following January, for it evidently prohibited the sale of some imported salt at auction. He was besought to suspend the act as unconstitutional so far as sales of merchandise imported from abroad were concerned. Executive Correspondence, Alabama Archives.

20. *Ordinances of the Virginia Constitutional Convention,* April-May, 1861, No. 48.

21. *Calendar of Virginia State Papers and Other Manuscripts from January 1, 1836, to April 15, 1869,* pp. 225-26. The governor's telegram

to the attorney-general would indicate that salt was being held by speculators at Lynchburg.

22. *The Richmond Enquirer,* Oct. 28, 1862.

23. She took her first steps in manufacture in Dec., 1861. See below, chapter VI.

24. Vance to Brown, Nov. 22, 1862, Vance Letter Book, Sep.-Nov., 1862, North Carolina Historical Commission; *O.W.R.,* series IV, volume II, p. 214.

25. *North Carolina Legislative Documents,* 1862-63, Nov. 17, 1862, Document No. 1, pp. 4-5.

26. *Public Laws of North Carolina,* 1862-63, Joint Resolution, pp. 53-54.

27. Gov. Vance renewed it Dec. 26 and April 13, 1863, Vance Letter Book, Sep.-Oct., 1862-63, pp. 79, 23-31.

28. T. W. Gore to Gov. Vance, Nov. 28, 1862, from Little River, South Carolina, Executive Papers, Z. B. Vance; E. Baum and Co. to Gov. Vance, Conneyboro Harry District, South Carolina, Nov. 26, 1862, *Ibid.*; T. D. Feaster to Gov. Vance, Nov. 20, 1862, *Ibid.*

29. John Dawson, mayor of Wilmington, to Gov. Vance, Nov. 13, 1862, *Ibid.*

It is interesting to note the immediate effect of an embargo in increased prices in adjacent states. One writer quotes a Judge Alin as remarking that he did not begrudge $10,000 for Governor Vance's order, as it made him $40,000 in Georgia. This occurs in a letter dated Wilmington, Nov. 26, 1862, the name of the addressee not appearing. *Ibid.*

30. Letter of Jonathan Worth to Gov. Vance, Nov. 26, 1862, *Ibid.*

31. George Reid to Gov. Vance, Petersburg, Virginia, Dec. 8, 1862, Executive Papers, Z. B. Vance, North Carolina Historical Commission.

32. Letter from James Carson to Gov. Vance, April 16, 1863, Z. B. Vance Executive Papers, April 1863; letter of J. M. Worth, Salt Commissioner, to Gov. Vance, April 22, 1863, *Ibid.*

A letter from D. G. Worth, at that time a private salt maker, to Jonathan Worth, his father, dated November 20, 1862, shows the first reaction to the embargo and may reflect the typical salt-manufacturer's view. He considered that embargo a blunder and noted the instant fall in price of salt to twelve dollars a bushel. "This act of the Gov. will shorten the production both in causing those who are already engaged in the business to relax their efforts and preventing others from going into it." *Correspondence of Jonathan Worth,* I, pp. 197-98.

33. Legislative Papers, 1862-63, North Carolina Historical Commission; *Public Laws of North Carolina,* 1862-63, Joint Resolution, Ratified December 8, 1862, pp. 52-53.

34. According to the statement of the governor in his annual message on November 25, 1862, *South Carolina House Journal,* 1862, Regular Session, p. 26.

35. *Ibid.,* Extra Session, April, 1863, p. 384.

36. *Louisiana Session Laws,* November-January Session, 1861, No. 42.

37. *O.W.R.,* series IV, volume II, pp. 250-51.

38. *Georgia Senate Journal,* Extra Session, 1861, p. 130. Its recommendation was embodied into Joint Resolution, No. 58, *Georgia Session Laws,* Extra Session, March-April, 1863.

39. *Georgia Senate Journal,* Extra Session, 1863, pp. 128-29.

40. Letter from Gov. Brown to the commanding officer, April 1, 1862, Governors' Letter Book, 1861-65, p. 240, Georgia Archives; *The Confederate Records of the State of Georgia,* II, p. 231.

41. *Session Laws of Georgia,* Regular Session, November-December, 1861, No. 1. It is worth while to note that the Continental Congress during the Revolutionary War urged subscription to the capital stock of private companies in order to secure salt.

42. Governors' Letter Books, 1861-65, p. 166, Georgia Archives. This contract was dated December 18, 1861.

43. See annual message November 6, 1862, *Georgia House Journal,* 1862, p. 28. Dr. Stotesbury began to produce in August, 1862, made about 300 bushels in all and displayed a sample to the legislators, *Ibid.,* p. 129.

44. *Georgia Session Laws,* Regular Session, November-December, 1862, No. 2.

45. *The Confederate Records of the State of Georgia,* II, p. 695.

46. Governor Clark Letter Book, 1861-62, p. 372, North Carolina Historical Commission.

47. *Ordinances of the Convention of the State of North Carolina,* Third Session, 1861-62, No. 6, ratified January 30, 1862.

48. This committee appears to have been especially helpless, and made no suggestion other than to offer advice to citizens that they revert to the habits of their ancestors, and then asked to be discharged from further consideration of the subject. Legislative Papers, 1862-63, North Carolina Historical Commission.

49. *Public Laws of North Carolina,* Adjourned Session, February, 1865, Joint Resolution, p. 32.

50. *Alabama Senate Journal,* Called Session, 1862, pp. 11–12.

51. *The Clarke County Democrat,* August 21, 1862.

52. *Alabama House Journal,* First Called Session, 1861, p. 25.

53. This bill was signed and became effective November 11, 1861, *Alabama Session Laws,* Regular Annual Session, 1861, No. 26. Alabama may have been consciously imitating the example set by the fathers in the Revolutionary War in the bounty offered on salt.

54. *Ibid.,* No. 27. This act was signed Nov. 19, 1861.

55. *Ibid.,* No. 28. Signed Dec. 7, 1861. *Alabama Session Laws,* Regular Annual Session, 1861.

56. Governor's Annual Message, *Alabama Senate Journal,* Called Session, 1862, p. 10.

57. Governor Shorter on August 2, 1862, requested Figh and Company to charge $1.25 per bushel at the wells, thus forcing the consumer and not the state to pay the bounty. Executive Correspondence, Alabama Archives.

58. A copy of the contract with Figh and Company is to be found in *Alabama Session Laws,* Regular Annual Session, 1862, No. 39. Meticulous directions were given by Governor Shorter for the salt to be turned over to the state agent, for the use of particular sacks, for the marking of the sacks or barrels containing it. See also the letter from Governor Shorter to John P. Figh, December 22, 1862, Executive Correspondence, Alabama Archives.

59. *South Carolina Session Laws,* December, 1861, Regular Session, No. 4598.

60. *Arkansas House Journal,* Regular Session, 1862, p. 46.

## CHAPTER VI

1. According to the statement of Governor Shorter in his annual message at the Called Session of October, 1862, *Senate Journal of Alabama,* p. 9.

2. *Ibid.,* p. 13.

3. To stimulate still further production for Alabama in Virginia, Governor Shorter offered to advance funds to the Tennessee company on condition of a bond to repay the loan with eight per cent interest in salt, computed at the rate of $1.75 a bushel, the salt to be delivered by the middle of January, 1863. A further condition stipulated that all salt, even beyond that turned over to the state, must be shipped to Alabama and sold at not more than five dollars a bushel for home consumption. Stated in a letter from Governor Shorter to General S. K. Rayburn, dated September 5, 1862, Executive Correspondence, Alabama Archives.

4. Governor Shorter did not hesitate to let the firm know that he was aware that after signing the Alabama contract, Mr. Jaques became interested as a co-partner with the Stuart, Buchanan Company so that the latter agreed to furnish him salt water at fifty cents instead of the 25 cents first charged. Letter of Governor Shorter to McClung, Jaques and Company of September 19, 1862. *Ibid.*

5. Governor's message, *Alabama Senate Journal,* Regular Session, 1863, p. 86.

6. The contract is to be found in Alabama Archives, Executive Correspondence. It is dated July 29, 1862. The Alabama Company had the right to finish up its contract for 50,000 bushels by the close of March, but the governor hoped that the company would expand, as Alabama could ship salt to the five northern counties cheaper from Virginia than it could from Clarke County.

7. The signed receipt is preserved in the Alabama Archives (No classification).

8. Shorter to the Honorable D. D. Avery, January 3, 1863, Executive Correspondence, Alabama Archives.

9. *Alabama Senate Journal,* Regular Session, 1863, p. 87. Governor Watts also made sundry small contracts for the manufacture of salt on the Florida coast, *Ibid.*

10. The details are to be found in the governor's message, *Alabama Senate Journal,* Called Session, November, 1862, p. 11. The problems of the agents are discussed in Chapter VII.

11. *Session Laws of Alabama,* Second Regular Annual Session, 1862, No. 38. The state also took over for $1,500 a lease of a private company in Clarke County with certain reservations, that of Dennis, English and Thomas.

12. See below Chapter IX.

13. The result of the quest in Alabama appears later.

14. Chien saw no hope except in seizure of the mine by the confederate

government, Pettus Papers, volume 58, Mississippi Archives. The report is dated September 16, 1862.

15. Governor Pettus made tremendous efforts to secure other boats to bring salt from Louisiana, appealing to General Pemberton for two boats in the Yazoo River, belonging to Mississippians, Pettus to Pemberton, October 21, 1862, Pettus Papers, volume 58. He also wrote Secretary of War Randolph on October 28, 1862, for prohibition of the seizure of four boats which he had engaged for the purpose, a request granted by Randolph, Pettus Papers, volumes 58, 60.

16. *O.W.R.*, series IV, volume II, p. 923; Pettus Papers, volume 58, Mississippi Archives. General Taylor condemned the seizure, but mere words were of little avail. See report of Pattison from New Iberia, dated November 18, 1862, *Ibid.* He condemned the lack of system and Judge Avery for not being able to forget his private interest in the mine. Pattison offered $5,000 rental for one of the pits for one month; Avery asked $50,000. Series G, 201, Mississippi Archives.

Governor Pettus asked the legislature to authorize him to impress a sufficient number of slaves to work the mine in Louisiana and also the necessary wagons and teams for transportation to the river, as the country appeared destitute of spades, shovels, wheel-barrows, carpenters, and tools, Annual Message, 1862. The desired permission was given by the legislature, *Mississippi Session Laws,* Called Session, 1862, chapter XIII, section I.

17. See reports of Dixon of September 22, 24, and of November 4, 1862. Pettus Papers, volume 58, Mississippi Archives. For the contract, see Series G, 201, Mississippi Archives.

18. *O.W.R.*, series IV, volume II, p. 923.

19. Such generosity was so unusual that Vardin wrote, "Such high-minded patriotism and liberality I met in no other instance." Vardin to Governor Pettus, September 20, 1862, Pettus Papers, volume 58, Mississippi Archives.

20. There was some timber on low land adjoining, which belonged to this same man. It is interesting that Vardin distrusted public works as "generally expensive from negligence and inattention", and so urged giving in part the privilege of salt-making to individuals who would obligate themselves to sell all salt made to citizens of Mississippi. *Ibid.*

21. *Mississippi Session Laws,* Called Session, 1862, Chapter XIII.

22. Turner had been so injured at Corinth as to necessitate his retirement from the army, but he was not debarred from such a post as salt agent, where his business experience prior to the war would count, Pettus Papers, series E, volume 59, Mississippi Archives.

The number of agents dispatched in frantic search of salt is confusing. Between Vardin and Turner a certain W. A. Strong seems to have served, for he reported on January 12, 1863, from Mobile on the fact that he could buy salt at fifteen dollars at the works, at seventeen in Mobile, and could ship out the salt bought from private wells very readily, *Ibid.*

23. The company asked an advance of $25,000 from the state, series G, No. 201, Mississippi Archives. The Norman Company was composed of citizens of Newton County, Mississippi. By the time they repudiated their contract, Turner had advanced $30,000-$40,000 in money and provisions.

See Turner's letter to Governor Pettus of November 9, 1863, series E, volume 61, Pettus Papers.

Turner's first reaction was distinctly against the state engaging in the manufacture of salt, as, though he found an abundance of salt water, "enough to make a sufficiency of salt for the whole Southern Confederacy," he found so great a scarcity of wood, such heavy investment necessary, and such high rentals for brine that it would be cheaper to buy the salt outright. It was his deliberate opinion that "we had better buy enough for present purposes and contract with private companies for what you will want. I think by furnishing the provisions to private companies and affording them the proper facilities that we can get as much salt as we want for 5 or 6 dollars,". His estimate of the cost of manufacture was six to eight dollars, exclusive of the $30,000 investment necessary to bore wells and to erect furnaces, Turner to Governor Pettus, February 24, 1863, Pettus Papers, series E, volume 59, Mississippi Archives.

24. See letter of Norman to Governor Pettus, dated August 15, 1863. Norman had had his collar bone broken in a railroad accident and was unable to attend to business properly, *Ibid.*, volume 61; also a letter from Turner to Governor Pettus of June 26, 1863, *Ibid.*

25. The Norman Company sold 1,000 bushels of salt which it had on hand in the open market at thirty dollars a bushel, an act which was severely criticized as not fair to the state, but Turner felt that it was justified, as fear of a Yankee raid had caused something of a panic, and he held it better for the company and for the state to sell the salt than to allow it to fall into the hands of the enemy. Turner to Governor Pettus, September 9, 1863. Turner himself did not escape without criticism. See letter of Thomas Brothers to Governor Clark, October 19, 1863. *Ibid.*

26. Series G, No. 201, Mississippi Archives.

27. *Ibid.*

28. Since the salt was received wet, it lost about one-fifth in drippage by the time it reached Atlanta from Saltville. The statement of terms is based on Governor Brown's message to the legislature and on his Letter Book. See *Georgia House Journal,* 1862, p. 28; Letter Books-Governors', 1861-65, p. 383, Georgia Archives.

It is worth noting that Senator Lewis's original proposal had been to manufacture on his personal account and to sell to the state for $10 per sack of three bushels; Governor Brown thought it better, because of their social relationship, to have the salt made on state account. *Ibid.* p. 302. In view of Governor Brown's later boast of unusually favorable terms from the Virginia proprietors, it is amusing that at first he considered the tariff of fifty cents a bushel "rather high, if not exorbitant". *Ibid.*

The contract may be found in Executive Department-Minutes, January 2, 1860-July 6, 1866, p. 395, Georgia Archives. See also a letter of Governor Brown to Secretary Randolph of June 9, 1862, asking that Bayliss Lewis be detailed to assist his father direct the work, *O.W.R.,* series IV, volume I, p. 1147.

29. To protect Temple for his time, it was stipulated in the contract that Lewis must take at least 60,000 bushels, even if the war terminated

before that amount had been made. Executive Department-Minutes, 1860-66, p. 395-96.

As no appropriation for purchasing salt had been made, Governor Brown met the situation by ordering the treasurer of the state railroad, the Western and Atlanta Railroad, to advance sufficient funds to the commissary general. He then brought the matter to the attention of the assembly in his next annual message in order that appropriations might be made to balance the accounts. *Georgia Senate Journal,* Regular Session, 1862, p. 30.

30. *Georgia House Journal,* Extra Session, April, 1863, p. 102. See also *The Confederate Records of the State of Georgia,* volume II, pp. 413-33 for the full transaction.

31. *Session Laws of Georgia,* Called Session, March-April, 1863, Joint Resolution, No. 58.

32. *The Confederate Records of the State of Georgia,* volume II, pp. 546-50.

33. *Georgia Senate Journal,* Regular Session, 1863, p. 194.

34. Executive Department-Minutes, 1860-66, p. 410, Georgia Archives. The governor informed the legislature that this firm proposed to devote their entire energies to importing salt from the Louisiana mine. *Georgia Senate Journal,* Regular Session, 1862, p. 30.

35. *The Confederate Records of the State of Georgia,* volume II, pp. 692-95.

36. *Session Laws of Arkansas,* Called Session, September-October, 1864, pp. 20-21.

37. On November 21, 1864, he reported in his annual message that he had paid for salt for distribution to soldiers' families $21,001.51, *Florida Senate Journal,* 1864, p. 26.

38. It will be noted that North Carolina is the only state which did not attempt purchase first.

39. *Ordinances of the State Convention of North Carolina,* Second Session, November-December, 1861, Nos. 8; Fourth Session, No. 18, ratified December 6, 1861, and May 9, 1862.

40. On December 9, 1861, Jonathan Worth expressed his surprise at the election of his brother, as salt commissioner, *The Correspondence of Jonathan Worth,* volume I, p. 159. Milton resigned the position, his resignation to take effect August 1, 1863, whereupon his nephew secured the post through the activity of Jonathan Worth. It should be noted that Daniel Worth was drawn from a firm engaged in the business of salt-making.

41. See letter of M. M. Worth to George Davis of March 18, 1862, *O.W.R.,* series I, volume LI, part 2, p. 506; *The Correspondence of Jonathan Worth,* volume I, p. 166; and Hamilton, *Reconstruction in North Carolina,* pp. 75-76. See map.

42. Letter Book, 1861-62, Governor Clark, p. 342, North Carolina Historical Commission.

43. *Ibid.,* p. 409. Governor Clark, like Governor Brown of Georgia, assumed power to appropriate money for that purpose, *Ibid.*

44. Letter of Governor Vance to J. M. Worth, October 1, 1862, Vance Letter Book, p. 862, North Carolina Historical Commission.

45. The rapid spread of the contagion is told by the following table:

| | | | | | |
|---|---|---|---|---|---|
| October 5 | there | were | 63 | new | cases |
| October 6 | " | " | 64 | " | " |
| October 7 | " | " | 58 | " | " |

See Hamilton, *Reconstruction in North Carolina,* p. 791.

The commissioner himself suffered in his own family, for his son, a youth of seventeen, was stricken and died in five days, Worth being his sole nurse, so that the commissioner was incapacitated for business for some days. After the boy's death, since the fever continued, he spent his time searching for materials needed in the work. This calamity is reported in a letter from Worth to Governor Vance of October 7, 1862, Vance Letter Book, p. 862, North Carolina Historical Commission.

46. Report of Daniel Worth to Governor Vance, September 16, 1863, Executive Papers, Z. B. Vance, North Carolina Historical Commission; report of Worth to Vance, November 7, 1863, *Ibid.* He lamented that he was obliged to let those men go home to recuperate, as there was no hospital connected with the works, Vance Letter Book, September-March, 1863-65, pp. 20, 269-70, North Carolina Historical Commission.

47. Where salt was exchanged for provisions—corn, fodder, bacon, etc., he computed the salt at the market price of fifteen to eighteen dollars a bushel. Worth to Governor Vance, September 16,. 1863, Vance Letter Book, 1862-63, September-October, p. 375.

48. He bewailed the fact that he had to pay five to six dollars per bushel for corn and then haul it fifteen to twenty miles. See report to Governor Vance of December 5, 1863, Vance Letter Book, December, 1863. For prices see his report to the governor of March 12, 1864, Executive Papers, Z. B. Vance, North Carolina Historical Commission.

As was entirely natural, Worth used some of his own salt to cure 30,000 pounds of pork for use at the works for the hands, which would, he thought, with the bacon on hand, suffice him for six to eight months. *Ibid.*

49. The "flats", as Worth termed them, were obviously flat barges used to convey great vats of brine and wood through the sound. See letter of Worth to Governor Vance, January 21, 1863, Vance Letter Book, 1862-63, September-October, p. 130, North Carolina Historical Commission.

The River Side Works, was the name by which Worth designated one of the works, as they were scattered at four different places on the coast.

50. Worth to Governor Vance, January 21, 1863, Executive Papers, Z. B. Vance, North Carolina Historical Commission.

51. One can readily sympathize with his complaint: "It is just impossible to get along where the military have control of everything." This was written as early as January 27, 1863, *The Correspondence of Jonathan Worth,* volume I, p. 228.

52. J. M. Worth to T. P. August, March 23, 1863, Executive Papers, Z. B. Vance, North Carolina Historical Commission.

53. Note should be taken of the fact that of 270 men, about forty were diseased and unfit for work most of the time. Vance Letter Book, September-October, 1863-65, pp. 186-87; letter of Worth to Colonel D. A. Barnes, June 6, 1864, Executive Papers, Z. B. Vance, North Carolina Historical

Commission. See also *North Carolina Legislative Documents,* 1863-64, Document No. 10, p. 6.

54. *O.W.R.,* series I volume LI, part 2, p. 840. General Whiting declared that southern newspapers reached union soldiers only one day old. See Whiting to Governor Vance, October 4, 1864, Z. B. Vance Papers, volume V, North Carolina Historical Commission.

55. See General Whiting's account to Governor Vance, April 22, 1864, Executive Papers, Z. B. Vance; and Worth's statement to the same person under the same date, *Legislative Documents,* 1863-64. Worth was confident after examination of the question that only one of the forty-seven men captured could justly be suspected of going over to the enemy willingly. The landing was too sudden to permit the salt-makers to escape. See also Worth's letters to Governor Vance of April 23 and 25; Governor Vance to Whiting, April 27; Governor Vance to Worth, April 27; and Whiting to Governor Vance, April 29. Executive Papers, Z. B. Vance, North Carolina Historical Commission; and *O.W.R.,* series I, volume LI, part 2, p. 881.

Whiting was not without those who shared his view. A citizen wrote Governor Vance from Fayetteville November 1, 1862, that free negroes should be sent to all the salt works for soldiers' wages, thus reducing the cost and releasing strong, young men for the army. This attitude is referred to elsewhere.

56. This order was dated June 6, 1864, *O.W.R.,* series I, volume LI, part 2, pp. 991-92; Whiting to Worth, June 6, 1864, Executive Papers, Z. B. Vance, North Carolina Historical Commission. See also Worth's report of April 25, 1864, to the governor, in regard to the federal raid, see Executive Papers, Z. B. Vance, North Carolina Historical Commission.

57. *O.W.R.,* series I, volume LI, part 2, p. 991. Under the same date, June 6, 1864, General Whiting reported his order to Adjutant-General Cooper, *Ibid.,* and to Governor Vance one day later, June 7, Whiting to Vance, Executive Papers, Z. B. Vance.

58. *The Correspondence of Jonathan Worth,* volume I, pp. 310-11, 315-16, 313-14, 318.

59. Governor Vance to Seddon, June 27, 1864, Executive Papers, Z. B. Vance, North Carolina Historical Commission.

60. *O.W.R.,* series I, volume LI, part 2, p. 1031; *The Correspondence of Jonathan Worth,* volume I, p. 319; Vance Letter Book, 1863-65, September-March, pp. 222-23.

61. *O.W.R.,* series I, volume XL, part 3, pp. 787, 789; see also Secretary Seddon to Governor Vance, July 21, 1864, Executive Papers, Z. B. Vance, North Carolina Historical Commission.

62. It would appear that detectives had been sent among the men to ferret out the disloyal, but with little success. The salt commissioner thought removal of the works costly and impracticable. Worth to Governor Vance, August 20, 1864, Executive Papers, Z. B. Vance, North Carolina Historical Commission. For the governor's alternative proposal see letter of Governor Vance to Worth, August 18, 1864, Vance Letter Book, 1863-65, September-March, p. 226. North Carolina Historical Commission.

63. Worth to Governor Vance, September 5, 1864, *Ibid.,* p. 241.

64. The telegram from General Bragg to Governor Vance reflects clearly General Whiting's influence, as it charges the very points on which the latter had harped so long, Bragg to Vance, October 25, 1864, Executive Papers, Z. B. Vance, North Carolina Historical Commission. See also letter of General Bragg to President Davis, October 25, 1864, *O.W.R.*, series I, volume XLII, part 3, p. 1171; and Whiting to Bragg, October 27, 1864, *Ibid.*, p. 1180.

65. *Ibid.*, p. 1214; Whiting to Worth, November 15, 1864, Vance Letter Book, 1863-65, September-March, p. 291.

66. Worth to Governor Vance, November 25, 1864, Executive Papers, Z. B. Vance; also letter of Worth to Vance dated November 15, 1864, Vance Letter Book, 1863-65, September-March, p. 290, North Carolina Historical Commission.

67. See the governor's annual message of November, 1864, *Legislative Documents*, 1864-65, Document No. I, pp. 14.

68. *Public Laws of North Carolina*, Regular Session, 1864-65, chapter 28.

69. *Ibid.*, Adjourned Session, February, 1865, Joint Resolutions, of a Public Nature, p. 32.

70. This invitation was issued to Governor Clark on June 13, 1862. It stated that if the governors declined the invitation, the proprietors would invite private enterprise to engage in the business. It is interesting that already their thought had fixed the form of payment as a tax upon each bushel of salt made. Letter Book, 1861-62, Governor Clark, p. 344-45, North Carolina Historical Commission.

71. The commissioners left Raleigh June 24, arriving in Saltville on the 27th. A letter from Charles Manly to Ex-Governor Swain, June 24, 1862, fixes the date of their departure, Swain Manuscripts, 1770-1869.

72. The contract is found in Letter Book, 1861-62, Governor Clark, pp. 360-61, North Carolina Historical Commission.

73. The commissioners felt that they could secure iron only through some government contract. In view of the pressing necessity for salt, they hoped for the benefit of 200 tons of iron from the government. The Richmond authorities granted the amount asked for, except, of course, gun metal. *Ibid.*, p. 360.

74. *Ibid.*, p. 404. Governor Clark also impressed on Governor Vance, who succeeded him in office on September 8, 1862, the necessity of urging on the work before it was stopped by winter weather.

75. *O.W.R.*, series IV, volume II, p. 182.

76. *Public Laws of North Carolina*, Regular Session, 1862-63, chapter 22.

77. Report of February 24, 1863, and of July 1, 1863, by Woodfin to Governor Vance, Executive Papers, Z. B. Vance, North Carolina Historical Commission.

78. The agent reports that North Carolina was the last state from which money was refused. He surmised that the object was to be paid in salt, since a number of private parties were paying three-fourths of all salt made as rent. Israel, sub-agent under Woodfin, to Governor Vance, September 2, 1863, *Ibid.*

79. *Ibid.*

80. The company condescended to accept $30,000 the day the agree-

ment was made, October 28, 1863, but only as a credit on the amount to be allowed by the arbitrators. The agreement is to be found in Executive Papers, Z. B. Vance under the above date.

81. For the contract, approved by the governor on April 9, 1864, see *Ibid.* and *Public Acts of North Carolina,* 1864-65, chapter 29.

82. Governor Vance wrote Woodfin on September 26, 1864, "I have been anxiously expecting you to go to Saltville for some months. Everything is going badly there judging from the tenor of my imformation. We cannot get a bushel shipped from the works though private parties continue to ship a great deal." Vance Letter Book, 1863-65, September-March, p. 260.

83. *Public Laws of North Carolina,* 1864-65, chapter 29. For the question of transportation of salt from Saltville, see below, chapter IX.

84. North Carolina was to detail eight to twelve agents to make the salt; allowance was fixed for fluctuation in the currency at the rate of $1.60 per gold dollar. The contract is found under the above date in Executive Papers, Z. B. Vance, North Carolina Historical Commission.

85. For the contract see *Ibid.*

86. The works were located at Little River, at the head of Murrill's Inlet, at Rikersville near Charlotte, on the coast of All Saints Parish, and one on Masonboro Sound, North Carolina, *Journal of the Convention of South Carolina,* 1860-62, Fourth Session, Appendix, pp. 686-87.

For the action of the Executive Council see *South Carolina House Journal,* Regular Session, 1862, p. 120.

87. Based on a statement in the governor's annual message, *South Carolina House Journal,* Regular Session, 1862, p. 26.

88. *Session Laws of South Carolina,* February-April Session, 1863, No. 4621.

89. Hilgard reports salt and soda as made by Governor Allen for Louisiana from water from pits dug in the Sabine bottoms two miles below Myrick's Ferry in northern Louisiana, *Supplementary and Final Report of a Geological Reconnoissance of the State of Louisiana,* p. 22. See also Veatch, *op. cit.,* p. 90.

90. *Session Laws of Texas,* Ninth Legislature, November Session, 1861, Joint Resolution, Chapter VIII.

91. It is interesting to note in this connection that the legislature desired the military commander of the department to dispose troops upon the northwestern boundary of the state so as to protect the line of travel for parties going to the Double Mountains to make salt. *Session Laws,* Tenth Legislature, 1863, Joint Resolution, Chapter VIII.

It appears that this action was based upon a reconnoissance of the salines on the northern frontier of the state made by a J. A. Bishop, Letter of Bishop to Governor F. P. Lubbock, October 9, 1863, Confederate Military Board Records, State Library Archives of the State of Texas.

92. See below Chapter IX.

93. The action of Clarke County was legalized after steps had been taken for county works. The commissioners were authorized to sell enough of their salt at market prices to cover the cost of the venture. Session Laws of Alabama, Regular Session, 1863, No. 221; see also

*The Clarke County Democrat,* November 27, 1862, when it reported progress.

94. Minute Book, Court of County Commissioners, Clarke County, War period. One hundred and fifty dollars was ordered paid over to a certain person to buy salt for indigent families before the county works were underway.

95. *Session Laws of Alabama,* Regular Annual Session, 1863, No. 97, section 3.

96. *The Richmond Enquirer,* September 26, 1862; letter of Colonel S. Forhan to Governor Vance, November 4, 1862, Executive Papers, Z. B. Vance, North Carolina Historical Commission. The minutes of the Shenandoah meeting are found in One Package of Papers Relating to Salt, 1860-65, Virginia Archives, (MSS.).

97. This statement is based on Governor Letcher's report to the assembly of Virginia at its session in January, 1862. *Virginia Documents,* 1862-63, *Governor's Message and Accompanying Documents,* Document No. 6, p. 6. This document is without special title.

98. *The Charleston Mercury,* July 31, 1862.

## CHAPTER VII

1. In Mississippi the quarter-master general discharged the duties of general salt agent in addition to his other tasks for several months until a separate salt agent was appointed. Quarter-Master General A. M. West served from April 13, 1863, to December 11, 1863, when the offices were separated and Z. A. Philips became general salt agent, serving through the rest of the war. Pettus Papers, series E, volume 68, Mississippi Archives; *Mississippi Senate Journal,* 1863, Called Session, Appendix, pp. 187-88; *O.W.R.,* series IV, volume II, p. 924.

2. The places in Georgia were Carterville, Atlanta, Columbus, Athens, Augusta, Griffin, Macon, Albany, and Savannah. *The Confederate Records of the State of Georgia,* volume II, p. 229. Governor Brown was willing to allow additional distributing centers if the people of a county at a public meeting appointed an agent who would agree to sell at the stipulated price with allowance only of freight charges additional. *Ibid.,* p. 230. At least one county secured an extra center. Letter Books-Governors', 1861-65, p. 333.

In Alabama, Tuscaloosa, Selma, Montgomery, Talladaga, Demopolis, and Eufaula were selected and the agents allowed a five per cent commission, not to exceed an annual income of $1,500. See letter from Governor Shorter to A. McKenzie, May 13, 1862, Executive Correspondence, Alabama Archives.

3. The legislature of Georgia at its regular session, November-December, 1862, authorized payment by the state of the cost of freight from the railroad to the different counties for salt designed for soldiers' families. It was apparently made retroactive, *Session Laws of Georgia,* 1862, No. 46.

4. These arrangements were worked out by Governor Brown for the first distribution in the summer of 1862. Executive Department-Minutes,

1860-66, p. 441, Georgia Archives; *Georgia Senate Journal,* Regular Session, 1862, pp. 28-29.

It was expected that state salt would be placed on the market as soon as the distribution was completed at prices to cover all costs so that purchasers financially able to pay for their salt would thus help pay for the donation to soldiers' widows. *Ibid.*, p. 29.

5. Letter Books-Governors', 1861-65, p. 390, Georgia Archives.

6. One was made in the summer and fall of 1862, one in the fall of 1863, and one, presumably, early in the fall of 1864. See *Georgia Senate Journal,* Regular Session, 1863, p. 30; *The Confederate Records of the State of Georgia,* volume II, pp. 728-32. Indigent refugee families were to be supplied gratis, but six dollars per bushel was to be paid out of the relief fund apportioned by the state to the county from which they came. *Ibid.*

7. See the several letters from the governor to J. P. Figh and Company, August 2, and September 1, 1862, and his letters to the agents at the various distributing centers. Executive Correspondence, Alabama Archives.

8. The governor set up a rule in the beginning to regulate the method by which the commissioner would fix the price, in the absence of legislative direction: "Estimate the daily expenses for labor and agents employed, and provision supplies consumed. To this sum add 20 pct. on gross amount of capital invested in the machinery, improvements, tools, wagons, teams, etc. Divide the aggregate amount thus obtained, by the quantity of salt made daily, and you will have the cost per bushel. To this cost you will add the cost of barrels, or sacking, as the case may be." Instructions of December 22, 1862. Executive Correspondence, Alabama Archives.

9. For the rule laid by the legislature for distribution and the governor's procedure and instruction to the agents concerned, see *Session Laws of Alabama,* Second Regular Annual Session, 1862, Nos. 38, 39; *Alabama Senate Journal,* Called Session, 1862, p. 12; and Governor's Message at the regular Annual Session, 1863, *Alabama Senate Journal,* 1863, p. 85-6.

10. Letter of Governor Shorter to Figh, September 1, 1862; Shorter to L. F. Johnson and Company, September 22, 1862; and Governor Shorter to the County Commissioners of Jackson County, July 15, 1862; and Governor Shorter to T. M. Gabbert, August 2, 1862, Executive Correspondence, Alabama Archives.

11. See a letter from Governor Shorter to Quarter-Master General Green, dated January 12, 1863. *Ibid.*

12. *Session Laws of Alabama,* Called Session, 1863, No. 6; Called Session, 1864, No. 6.

13. *Session Laws of Mississippi,* Regular Session, 1863, chapter XLV; *Session Laws of Arkansas,* Called Session, 1864, pp. 20-21, ratified October 1, 1864.

14. The law which made the above machinery for distribution effective in Mississippi was not passed until December, 1863. *Session Laws of Mississippi,* Regular Session, 1863, chapter XLV.

15. *Mississippi Senate Journal,* Called Session, 1862, and Regular Ses-

son, 1863, Appendix, pp. 188-89; *Session Laws of Mississippi,* Called Session, March-April, 1864, chapter IV.

16. The evidence of the second distribution comes from a receipt from H. F. Shelton to Z. A. Philips, dated April 15, 1865: "Received of Z. A. Philips, general salt agent, State of Mississippi, 7210 lbs. of Salt to pro rata share due Rankin County, Miss. on the 2d Distribution for which I have this day paid him $1137.50." Series G, No. 201, Mississippi Archives.

17. *Public Laws of North Carolina,* Adjourned Session, 1862-63, Joint Resolution, pp. 73-4; J. E. Fenton to Governor Vance, January 9, 1863, Executive Papers, Z. B. Vance, North Carolina Historical Commission.

Sometimes the inequality in the amount was due to the fact that the salt agent had to give preference to counties which had furnished labor or supplies. It varied, for instance, in July, 1863, from 300 bushels to Clay County up to 3,710 to Franklin County. See Woodfin's report to Governor Vance, July 1, 1863. *Ibid.*

18. Jonathan Worth, state treasurer, wrote very sharply to his brother, the salt commissioner, in May, 1863: *"Salt should in no instance be delivered to County agents until paid for.* This is distinctly provided for in the first section of the Salt ordinance. If you have been disregarding this provision of the ordinance you should at once notify each County Comr. that none will in future be delivered except upon payment, as required by the ordinance, and none will be delivered to counties in arrears until all arrearages are paid." *The Correspondence of Jonathan Worth,* volume I, p. 236.

19. The whole story of the handling of the salt question in Virginia is related in chapter IX.

20. *Virginia Documents,* 1862-63, Document No. 6, pp. 7-9. This document, bearing no title, is bound with *Virginia Senate Journal and Documents,* 1862-63. The refugees were included by an amendment to the law passed September 18, 1863. In the winter of 1862 the counties were receiving large consignments of salt, ranging from 40,000 to 86,000 bushels. See One Package of Papers Relating to Salt, 1860-65, Virginia Archives.

21. *Session Laws of Virginia,* Adjourned Session, 1863, chapter 17. An amendment, passed in the session of 1863-64, similar to the provision in Mississippi, allowed disposal of the salt other than by quota distribution to prevent its falling into the hands of the enemy, or if for any reason it proved impossible to distribute it to the people of the county. *Ibid.,* Session of 1863-64, chapter 7, section 2.

22. Wise, *History of the Seventeenth Virginia Infantry,* p. 170.

23. Dodge, "Domestic Economy in the Confederacy", *Atlantic Monthly,* volume 58, p. 231.

24. A partial list of the Mississippi agents is illuminating:

A. M. West, General Salt Agent, April 3-November 1, 1863.

Z. A. Philips, General Salt Agent, November 1, 1863-October, 1865.

D. S. Pattison, sent to the New Iberia Mine to buy salt, 1862.

Thomas E. Helm, purchasing agent for cotton and salt.

O. H. Weston, Assistant salt agent, January 1, 1864.

T. J. Arnold, Assistant salt agent, February-June, 1864.

Dunning, Assistant salt agent, 1864.

J. H. Paulett, Overseer, March-June, 1864.

L. D. Rhodes, Assistant Salt Agent, December, 1863.

T. Lomax, Assistant Salt Agent.

W. C. Turner, Assistant Salt Agent, October, 1863-

Compiled from papers in series G, No. 201, Mississippi Archives; and *Mississippi Senate Journal,* Called Session, 1862, Appendix, p. 187.

A similar list could be readily compiled for Alabama or Georgia.

25. The large group of assistants under Superintendent Clarkson at the Virginia state works is perhaps the most interesting. It included the following:

1 Assistant Superintendent at a salary of $2000 a year.

1 Clerk at a salary of $2000 a year.

1 Assistant Clerk at a salary of $2000 a year.

1 Shipping Clerk at a salary of $1500 a year.

3 Purchasing Agents at a salary of $2000 a year.

1 General Superintendent of Wood Cuttings at a salary of $2000 a year.

3 Overseers of Choppers at a salary of $100 a month.

1 Wagon Master at Wood Lot at a salary of $100 a month.

1 Wagon Master at Furnaces at a salary of $100 a month.

2 Managers at the Upper Furnaces at a salary of $—

"Answer of Superintendent J. N. Clarkson Relative to the Condition of the Salt Works", (Dated February 24, 1864), *Virginia Documents,* 1863-64, Document No. 29, p. 4.

26. Governor Shorter was criticized for devoting so much time to salt which might better, it was held, have been given to procuring arms. After wading through the sheaf of correspondence on the subject, one realizes, at least, the grounds of the complaint.

Appointed in the summer of 1862, McGehee was continued in the post of general director of the salt works by the legislative act of December following.

27. See the governor's instructions of December 22, 1862, Executive Correspondence, Alabama Archives; *Session Laws of Alabama,* Fourth Regular Annual Session, 1864, No. 87. Until the governor could secure an agent to be resident at the Lower Reservation to receive the salt from the Figh Company, McGehee was obliged to handle that also. When the salt agent was appointed, he was placed under the supervision and control of the quarter-master general. See letter of Governor Shorter to Green, March 21, 1863, Executive Correspondence, Alabama Archives.

For Governor Pettus's instructions of April 13, 1863, to A. M. West, although he was not yet general state agent, see Pettus Papers, volume 60, Mississippi Archives.

28. The Joint Committee on Salt Supply for Georgia, reporting April 13, 1863, found that it was practicable to make iron in that state and that the people of several counties were embarking on the work for their own supply. But in the event of absolute need, it felt confident that the confederate government would permit the use of as much iron from its contracts as would be needed to make the kettles. *Georgia Senate Journal,* Extra Session, 1863, p. 129.

Governor Shorter declared that if the state foundries could not cast the kettles and pans, the people were ruined, and hoped that all pig-iron which was not suitable for gun-metal would be diverted for this purpose. Governor Shorter to General C. I. McRae, September 17, 1862, Executive Correspondence, Alabama Archives.

29. See Woodfin's report to Governor Vance, November 27, 1862, *North Carolina Legislative Documents,* 1862-63, Document No. 13, p. 3.

30. See letter of Israel to Captain Dodd, September 8, 1863, Civil War Papers, Quarter-Master's Department, May-September, 1863, Unpublished papers of the War Department; Israel to Governor Vance, April 25, 1864, Executive Papers, Z. B. Vance, North Carolina Historical Commission. He used some of the material to clothe the hired negroes whose owners failed to provide them.

Naturally, the rush to get into the salt business in the fall of 1862 led to the most extravagant prices for necessary materials of all sorts, a factor which contributed to the worries of the agents. See J. M. Worth's first report, September 19, 1862, Vance Letter Book, 1862, pp. 8-10, North Carolina Historical Commission.

31. Daniel Worth to Governor Vance, April 12, 1864, Executive Papers, Z. B. Vance, North Carolina Historical Commission.

32. "Report of the Joint Committee on Salt in Respect to Salt for Georgia", *Virginia Documents,* Extra Session, 1863, Document No. 10, p. 17.

33. The first salt commissioner of North Carolina wrote his brother, Jonathan, from Wilmington, May 16, 1863, "I was all the time annoyed to death on acct. of a short allowance of provisions and I was really alarmed about getting it—and I sent out agents with directions to buy." *The Correspondence of Jonathan Worth,* volume I, pp. 237-38.

34. Woodfin to Ex-Governor Swain, November 17, 1862, Walter Clark Manuscripts, volume IV, North Carolina Historical Commission. Slaughtering the animals for food presented the additional problem of disposal of the hides and tallow, which, however, paid for far more than half the initial cost of the animals for meat. *Ibid.*

35. See Daniel Worth's report to Governor Vance, December 5, 1863, Worth to Vance, February 18, 1864, Executive Papers, Z. B. Vance, North Carolina Historical Commission.

36. Woodfin contracted with this man for at least 1,000 bushels of corn, 200 bushels of wheat and oats, eight tons of hay, thirty head of young beef, and pasturage for twenty other head of cattle. He made a similar contract with John Shannon, the collector of revenue for Smyth County. These contracts were reported in his letter of March 31, 1865, to the governor. *Ibid.* The letter is also to be found in "Report of the Joint Committee on Salt in Respect to Salt for Georgia", *Virginia Documents,* Extra Session, 1863, Document, No. 10, p. 13.

37. *The Correspondence of Jonathan Worth,* volume 1, pp. 237-38; Report of Worth to Governor Vance, July 2, 1863, Executive Papers, Z. B. Vance, North Carolina Historical Commission.

38. See Woodfin's letter to Governor Vance, January 3, 1863, *Ibid.* He heard of men promising one-third, or even one-half, of the salt for hauling it. Such contracts, it should be observed, would defeat the very purpose of state activity in the manufacture of salt. Letter of July 1, 1863, *Ibid.*

39. Woodfin to Governor Vance, February 24, and July 1, 1863, *Ibid.*

40. Statement of Superintendent Clarkson, "Contract Entered into between the Joint Committee of Senate and House of Delegates and Stuart, Buchanan and Company in Relation to a Supply of Salt", *Virginia Documents,* 1862-63, Document No. 43, p. 153.

41. Of his own initiative Worth arranged for about a hundred hands to go home in May, 1863, for twenty to thirty days in order to gather their crops. Worth to Governor Vance, May 26, 1863, Executive Papers, Z. B. Vance, North Carolina Historical Commission.

In 1863 Superintendent Clarkson was sending to Tennessee to purchase teams to haul and to Georgia for sacking. Report of Clarkson, "Report of the Joint Committee on Salt in Respect to Salt for Georgia", *Virginia Documents,* Extra Session, 1863, Document No. 10, p. 12.

42. Woodfin to Governor Vance, July 1, 1863, and January 3, 1863, Executive Papers, Z. B. Vance, North Carolina Historical Commission.

43. Worth to Governor Vance, March 12, 1864. *Ibid.* He reported that at least three-fourths of the private works on the coast had long since suspended by May 6, 1864, because of the scarcity of labor and supplies and the long distance of the haul for the wood. Vance Letter Book, 1863-65, September-October, pp. 186-87, North Carolina Historical Commission.

44. *Ibid.,* pp. 269-70.

45. Letter of Worth to Vance, September 19, 1862, Vance Letter Book, 1862-63, pp. 8-10, North Carolina Historical Commission.

46. Report of Woodfin to Governor Vance, November 27, 1862, *North Carolina Legislative Documents,* 1862-63, Document No. 13, p. 2.

47. *Session Laws of Alabama,* Second Regular Annual Session, 1862, No. 69, Ratified November 28, 1862.

48. The military officer readily agreed to detail them. See letter of Woodfin to Governor Vance, February 26, 1863, and Vance's reply of March 4, 1863. Vance Letter Book, 1862-63, September-October, pp. 163-64, North Carolina Historical Commission.

49. Worth to Governor Vance, April 12, 1864, Executive Papers, Z. B. Vance, North Carolina Historical Commission.

50. *Ibid.* A little earlier than the date given above, on October 15, 1864, Governor Vance had interposed against conscription of the directors and hands of the Stuart, Buchanan Company. *Ibid.*

51. The contract was originally dated March 23, 1864, and may be found in Executive Papers, Z. B. Vance. See also Woodfin's report to Vance of March 21, 1865, *Ibid.*

52. See Woodfin to Governor Vance, February 24, 1863, *Ibid.* Woodfin insisted that the most distant counties must have the preference to what railroad service was available. For evidence that other states were busy securing teams for long distance hauling see A. K. Ferrar to Governor Pettus, February 16, 1863; T. Lomax to Pettus, January 19, 1863; W. L. Dogan to Pettus, February 9, 1863, Pettus Papers, series E, volume 59, Mississippi Archives. That Governor Shorter encouraged Alabamians to make the long haul from Virginia may be seen in a letter by him to R. B. Kyle, November 24, 1862, Executive Correspondence, Alabama Archives.

53. For Governor Vance's reply to Woodfin, dated March 4, 1863, see

Vance Letter Book, 1862-65, September-October, p. 164, North Carolina Historical Commission.

54. Executive Papers, Z. B. Vance, North Carolina Historical Commission.

55. "Documents Relative to Salt," *Virginia Documents,* No. 43, pp. 137-38. A further problem in connection with assessment proceedings was the fact that the owner always selected a partisan bitterly opposed to the policy of state manufacture of salt, who by refusing to agree to any umpire to adjudicate between the assessor appointed by the proprietors and the one selected by the state, could indefinitely postpone the assessment or compel the selection of an umpire favorable to his view, thus practically dictating the award. Nearly all the assessors likely to be chosen by the proprietors had some interest in the award by virtue of having property likely to be impressed. *Ibid.*

56. See a letter by Thomas Brothers to Governor Pettus, dated October 19, 1863, Pettus Papers, series E, volume 61, Mississippi Archives.

57. For complaints against Woodfin to the governor see letters of G. W. Nicholson, May 2, 1864, of B. C. Washburne, April 30, 1863; for Woodfin's defense, see his letter to Governor Vance of January 3, 1863, all found in Executive Papers, Z. B. Vance, North Carolina Historical Commission.

The county agents in their turn did not escape. See Colonel West's complaint concerning a man of Holmes County, Mississippi, in a letter to Governor Pettus, dated February 10, 1863, Pettus Papers, series E, volume 59, Mississippi Archives. A complaint against the agent of Burke County, North Carolina, is recorded in a letter by S. C. Wilson to Governor Vance, of December 8, 1862, Executive Papers, Z. B. Vance, North Carolina Historical Commission.

58. "Documents Relative to Salt," *Virginia Documents,* No. 43, p. 173.

59. Testimony both favorable and adverse was heard. See *Ibid.,* pp. 143-202. One man declared, "There is much rottenness in Saltville, I tell you.", p. 172. For Clarkson's own statement see *Ibid.,* pp. 164-65.

60. The report referred to in the text is that of November 27, 1862, to which the writer has had occasion to refer a number of times, and which was valuable despite the difficulties under which it was drawn up. *North Carolina Legislative Documents,* 1862-63, Document No. 13, p. 1.

John Milton Worth evidently found the reports so burdensome that he failed to submit them properly and regularly, as is evident from comments by his brother. See *The Correspondence of Jonathan Worth,* volume I, pp. 262-63. Clarkson was not invulnerable in this regard, despite his copious and lengthy reports. See "Documents Relative to Salt," *Virginia Documents,* No. 43, p. 158.

For an illuminating statement as to the problems besetting a salt agent or superintendent, see Clarkson's report to the Virginia Board of Supervisors, dated September 4, 1863, *Virginia Documents,* 1862-63, No. 6 (Without special title); and Document No. 43, cited above.

61. Davis, *The Civil War and Reconstruction in Florida,* p. 205.

62. Daniel Worth reported 460 names on his roll when he took charge, but he held that 190 had died or been discharged or taken by the enemy. Some were always unavoidably absent, so that he averaged around 270

men. Vance Letter Book, 1863-65, September-March, p. 187, North Carolina Historical Commission.

63. See letter of W. M. Poisson to Governor Vance, November 1, 1862, Executive Papers, Z. B. Vance, North Carolina Historical Commission.

64. See Philips's reply to Governor Clark, May 22, 1864, Pettus Papers, series E, volume 65, Mississippi Archives: Worth to Vance, September 19, 1862, Vance Letter Books, 1862-63, September-October, p. 9.

## CHAPTER VIII

1. See map. On both sides of Jackson Creek there are a number of artesian wells still flowing in 1930, and the crumbled remains of hundreds of wells. In a very few places the earth works to protect the salt-wells may still be traced.

2. This same scene of wagons pressing in from the entire region within a radius of hundreds of miles would have been presented at any of the major wells. At the Louisiana mine, for instance, it was not at all unusual to find 100 to 200 wagons lined up, awaiting their turn to secure a supply of salt.

Woodfin reported on November 27, 1862, concerning conditions at Saltville, "The wagons are crowding on us, and are being loaded at the rate of twenty-three to twenty-eight per day, for the last week or two especially; and they are now crowding in so closely on each other, that they cannot be loaded without longer delay than is desirable; and I fear that as the stream is increasing, they must soon be in each other's way." *North Carolina Legislative Documents,* 1862-63, Document No. 13, p. 7. Such crowding, reported as early as 1861, resulted in wagons waiting two days or more before they could be loaded.

3. Captain J. E. Kennedy, relating his experience, stated that he took six or eight of the iron pots used on his plantation for scalding the hair off of his hogs at packing season in which to make the salt. *Advertiser,* (Montgomery), August, 1901. In North Louisiana they often used the round sugar kettle brought up from the sugar mills, below Alexandria, though the long, shallow pans thirty feet long by eight feet wide, were also to be encountered there. A peculiar, sugar-loaf shaped kettle was cast at Alexandria in 1863. Huge steam-boilers, sawed in half, were often utilized at all the works.

4. The residents of the section dreaded the crowds for another reason: among the negroes were always some who would steal chickens, hogs, and corn.

5. Where a man had come to procure a small supply by barter for private family use rather than for a large plantation, he at once set about exchanging the corn or provisions which he had brought at the stores or with the individual salt-makers.

6. This last statement, made at the time of the Clarke County Centennial, is based on the oral statement to Mr. Mathews of Dr. I. J. Krouse of Suggsville, Ala., who worked at the Central Works all the time they were in operation. He added, as evidence of intimate acquaintanceship with the first efforts, that his neighbor, J. L. Jeffries, and his elder brother put up the first furnaces to make salt for their own use.

They afterwards enlarged their furnaces and made the first salt offered for sale. Before long the demand was greater than the supply. For these intimate details from men who lived at the works the writer is indebted to a manuscript by Mr. T. L. Head, written in 1910, which is preserved in the department of Archives and History of Alabama, A History of Clarke County.

Jeffries put up what he considered a large furnace of eight or ten wash-pots and of three or four large kettles which he bought in the neighborhood so that during the first year of the war he was making from eight to ten bushels of salt a day. *Ibid.*

7. J. C. Townsend, who is quoted by M. M. Mathews, in his manuscript work on A History of Clarke County, made quick work of hauling, for he cut his wood on the hill and slid it down to his furnace, which had been built at the foot of the hill.

Kennedy in his account of his salt-making activities says that he arranged his big pots in a row so that fires could be built under them.

8. *Ibid.* The Mathews manuscript.

9. In Louisiana, something of the same sort occurred. At the Rayburn Salt Works a toll or rental of 37½ cents a bushel was collected for the use of the brine. It was claimed that at the peak of the operations Rayburn collected as much as $375 a day, though old residents thought this statement extravagant. The writer is indebted for this opinion to J. P. Whittington of Alexandria, Louisiana. See also: Veatch, *op. cit.,* 72-73.

10. It is possible to follow this old tram-road even today; where the road circles around the ridges, piles of rocks are to be seen which were deposited there for transportation by the tram-car to the works. It was said that after the war a fence nearly a mile long was built from rocks and bricks of the old furnaces by a certain Pink Kelly. The Mathews Manuscript.

11. The salinity of the brine depends on the season, being less shortly after the flood time, more late in the summer.

12. Some of these tall chimneys remained standing a long time, but all have now tumbled into ruins.

13. I am indebted for this detailed description of the construction of the works to Mathews's manuscript. He, in turn, based it on the oral accounts of men who had participated in the construction at that time. The Louisiana wells were protected by earth-works or levees, or platforms. The brine was conveyed to the furnaces by troughs supported on forked poles. Veatch, *op. cit.,* p. 85.

14. Mathews states that at least thirteen men and firms operated there. *Op. Cit.*

15. The records have not been as well preserved for the Upper Works as for the others.

16. Mathews, *op. cit.*

17. Just as scruples against fighting on Sunday yielded in the presence of war necessities, so scruples against industrial enterprises being conducted on Sunday gave way to war exigencies. Note Woodfin's reluctance to make salt on Sunday on November 17, 1862, when he anxiously sought counsel of ex-Governor Swain on this "question of morality," and declared that at first he would not entertain the proposition. Walter Clark MSS., volume IV, North Carolina Historical Commission. Evidently his

conscience troubled him less as soon as he was able to get enough brine for seven days' boiling a week, for on November 27, we find him reporting to Governor Vance as follows:

"Seeing the great necessity that is upon us, and that every day's boiling is more than enough to save one hundred thousand pounds of pork, and that to cool off on Sunday involves the loss of more than half of Monday to get up the heat, etc., and after mature reflection and free consultation with friends whose opinions I value, I came to the conclusion to run the furnaces on Sunday, and to this all my assistants and associates in this work are fully agreed, including the negroes, who are eager to receive the wages." *North Carolina Legislative Documents,* 1862-1863, No. 13, p. 6.

18. Dr. T. J. Prim's name has come down as such a medical attendant.

19. At the time that Mathews wrote his manuscript in 1923, several old salt-makers even remembered a Methodist minister's name, J. H. Ewing, as that of one clergyman who visited the works in 1862. He may have been the man sent by the conference.

20. This story is told by Fleming, *Civil War and Reconstruction in Alabama,* p. 160, note 5. Any one who follows Woodfin's reports, will note the delicate youth, Hawkins, often alluded to.

21. Mathews, *op. cit.*

## CHAPTER IX

1. *Session Laws of Arkansas,* Regular Session, 1862, Joint Resolution, p. 84.

2. The various state Journals, minutes of legislative committees, and legislative papers yield interesting data on this head.

3. *Florida Session Laws,* 1862, Joint Resolution, No. 13. Such privilege had been sought of Florida some months earlier. In April preceding Governor Shorter had requested such a privilege of Governor Milton for Alabamians, as the salines of Alabama were not sufficient for her needs. The request was granted. See letter of Milton to Governor Shorter, May 13, 1862, series W, Alabama Archives.

The legislature of South Carolina expressed appreciation of the Florida offer in most cordial terms, and urged the citizens to avail themselves of the offer. *Reports and Resolutions of the State of South Carolina,* Regular Session, 1862-63, pp. 343-44.

4. This offer is indicated in a letter of acknowledgment by Governor Brown, dated April 21, 1862. The writer is not implying failure of delivery; the records are merely silent on the subject. Letter Books-Governors', 1861-65, p. 269, Georgia Archives.

5. *Alabama Senate Journal,* Called Session, November-December, 1862, p. 12; *Session Laws of Alabama,* Regular Session, 1862, No. 40. Governor Shorter's proclamation extending such an invitation to the Alabama salt region is dated September 9, 1862. See Governor Shorter to Governor Pettus, January 3, 1863, Pettus Papers, series E, volume 59, Mississippi Archives.

Under the surface there was, however, some bitterness on the part of the states unfortunately placed with regard to salines. Note the comment of Pattison to Governor Pettus on November 18, 1862, in regard to his search for salt. "I will keep trying, but I assure you it is painful, as the Agent of a great state whose sons are pouring out their blood like water to defend the country, to become a mendicant to any man for one of the first necessities of life, for the wives and families of these soldiers, who may be said to be defending this very property." *Ibid.*, volume 58.

6. Woolsey to Philips, May 2, 1864, *Ibid.*, series E, volume 65.

7. A committee to confer with the lessees was created December 13, 1861; a little later a Joint Committee to visit the salt works was appointed to see if the capacity of the works could be increased; and it was further instructed to consider bounties and the manufacture of salt from sea water. *Virginia House Journal,* Session of 1861-62, pp. 36, 37-38.

The charge was brought before the committee that the lessees were making an enormous profit, that they sold the salt at 75 cents a bushel when it cost them only 30 cents, including the rental of the lease. It was felt that they were unwilling to make large expenditures to increase the works for fear the blockade might be lifted. "Minority Report of the Joint Committee to Correspond with the Lessees of the Salt Works," *Virginia Documents,* 1861-62, Document No. 47, p. 3.

8. See the following letter of Stuart, Buchanan Company to the legislative committee, dated May 5, 1862. It is interesting to encounter here a repercussion of the agitation which Thomassy had set up over solar evaporation of salt. "We would then allow with pleasure the state to put up any erection either to boil salt water, or evaporate it by the sun. To put up the first, as they would be of no service to us at the termination of the war, we think a moderate compensation for the use of the water should be allowed us. But if solar works are put up, we would ask nothing more for the water than the right to use those works from the opening of the blockade to the end of our lease. In either case, we think it should be provided that the state should cease to manufacture salt on the opening of the blockade." "Report Relative to the Purchase of Salt", *Virginia Legislative Documents,* 1861-62, Document No. III, p. 4.

9. "Report of the Special Joint Committee Relative to Supply of Salt", *Virginia Legislative Documents,* 1861-62, Document No. 16, p. 3; *Session Laws of Virginia,* 1861-62, Joint Resolution, No. 36; *Session Laws of Virginia,* Special Session, April-May, 1862, Joint Resolution, No. 22.

10. *Ibid.*, chapter 16.

11. The lessees were willing to transfer their lease to the Preston and King estates, which had still seven years to run, to the state of Virginia for $300,000. "Minority Report of the Joint Committee Appointed to Correspond with the Lessees of the Salt Works", *Virginia Legislative Documents,* 1861-62, Document No. 47, p. 4. They paid a rental of $20,000 for the three wells of the King estate, $30,000 rental for the Preston estate.

12. For the opinion of the attorney general see Report of J. R. Tucker to Governor Letcher under date of June 20, 1862, *Calendar of Virginia State Papers and Other Manuscripts from January 1, 1836, to April 15, 1869,* pp. 210-11.

13. For the contract see *Virginia Legislative Documents,* 1862-63, Document No. 3, p. 3. This document has no title, but is bound with the governor's message and is dated August 5, 1862.

14. *Virginia House Journal,* Called Session, September 1862, Document No. I, p. V. Governor Letcher felt that the necessary appropriation for manufacture by the state should have been made at the preceding session, but that even in September, 1862, salt works could be erected and the meat saved for the season. *Ibid.,* pp. III-IV.

15. The work of the Special Joint Committee on Salt Supply, which sat from September 17 to 23, 1862, is worthy of passing comment. It was greatly interested in securing a supply of salt from Kanawha, then in possession of the confederacy. It debated seriously getting possession of the Virginia works by purchase or lease. A representative of the Stuart, Buchanan Company was summoned before the committee and sharply questioned in regard to the supply of brine, the cost of production, and the cost of the works. The committee was also instructed to consider experimental borings in Pittsylvania County. One Package of Papers Relating to Salt, 1860-65, (MSS.) Notes of Proceedings of the Joint Committee on Salt Supply. Virginia Archives.

16. For the entire act see *Session Laws of Virginia,* Called Session September-October, 1862, Chapter I. This act was approved October 1.

17. The reader is reminded that sixty counties of Virginia had, during the period from April 11 to August 13, 1862, negotiated contracts with the Saltville proprietors for salt. No deliveries up to October exceeded 3,000 bushels, except to Richmond city. *Virginia Legislative Documents,* 1862-63, Document No. 1, pp. 17-18. See above p. 110.

18. *Virginia House Journal,* Adjourned Session, January-March, 1863, Document No. 1, Message of the Governor, January 7, 1863, pp. IX-XIII. For the contract with Charles Scott, Stuart, and Palmer, see *Virginia Legislative Documents,* 1862-63, Document No. 6, pp. 3-4. The contract is dated October 22, 1862.

19. For the full account of Governor Letcher's efforts between October 8 and January 1, see his message to the legislature on January 7, 1863, *Virginia House Journal,* Adjourned Session, January-March, 1863, Document No. 1, pp. IX-XIII. See also the report of the quarter-master general, Smoot, and of Colonel French in regard to the Kanawha effort, *Virginia Legislative Documents,* 1862-63, Document No. 6, pp. 10-34.

20. *Virginia House Journal,* Adjourned Session, January-March, 1863, p. 41. The assembly at its session in September, 1862, had created a committee to investigate Stuart, Buchanan and Company. This body exonerated the company of selling to extortioners in a report submitted September 29, 1862. *Virginia Legislative Documents,* Document No. 3. This document is without title and is bound as an appendix to the governor's message.

21. The lease was to run from April 1, 1863, to April 1, 1864. The ten furnaces included the four known as the "River Works", and counted the double furnaces as two; the state had the right to cut and quarry equally with the original lessees from the King and Preston estates. The company was bound to furnish to Virginia brine to keep the furnaces going at their full capacity prior to any other person or company, including themselves; and the lessees agreed to sell one-half their wagons, provisions, and half of all wood then cut; they further agreed to deliver half of Superintendent Clarkson's slaves which he had hired them. For the contract in full see "Report of the Joint Committee on Salt,"

*Virginia Legislative Documents,* 1862-63, Document No. 31 (dated March 27, 1863).

22. These four furnaces were leased to men named Clarkson, Friend, Kelley, and Gardner.

23. *Session Laws of Virginia,* Adjourned Session, 1863, chapter 17. The superintendent was, as a matter of course, heavily bonded for $200,000; he was allowed one deputy manager and a clerk as assistants, a number which had to be later increased.

See *Virginia Legislative Documents,* 1862-63, Document No. 37, for a contract negotiated March 6, 1863, to lease the wells to Clarkson for $100,000 in salt after impressing them from the original lessees. He was to deliver 700,000 bushels of salt in eight months after March 15, 1863, at $1.50 a bushel. This plan was apparently dropped. The minority of the committee had still another plan for purchase of salt supply to the amount of 700,000 bushels from Charles Scott and Company, another company operating at Saltville. *Ibid.,* Document No. 38.

24. *Session Laws of Virginia,* Adjourned Session, February-March, 1863, chapter 18; "Report of the Joint Committee on Salt in Respect to Salt for Georgia", *Virginia Legislative Documents,* 1863, Document No. 10, p. 12.

The Georgia supply was on one occasion subjected to inspection, and when she refused to comply with the law, the case was presented to the grand jury of Smyth County, but it was finally ignored by the jury, as the planters from the surrounding country protested. See "Communication from the Governor of Georgia Relative to Supply of Salt, *Virginia Legislative Documents,* Document No. 6, pp. 5-6.

25. The sum finally paid by the state of Virginia for the lease was $388,187, $300,000 for the brine and realty, the rest for personality. *Virginia Legislative Documents,* September-October Session, 1863, Document No. 10, p. 16.

The committee states that the works were turned over to Virginia on June 8, but Clarkson, who should know, gives June 7.

26. "Report of the Joint Committee on Salt in Respect to Salt for Georgia", *Virginia Legislative Documents,* Document No. 10, p. 16.

27. *Ibid.*

28. "Report of the Joint Committee on Salt to the General Assembly", *Virginia Legislative Documents,* Document No. 9, p. 3.

For Clarkson's suggestion to impress the Preston well, see *Virginia Legislative Documents,* Extra Session, 1863, Document No. 10, p. 11. For evidence that Clarkson felt that the Stuart, Buchanan Company was hampering the works, see *Ibid.,* p. 16. It will be recalled that under the law of March 30, 1863, he had power to impress, if the company failed in its contract.

The Joint Committee on Salt considered at length the causes of delay in delivery of brine and reported on October 28, 1863, and again later in the next regular session on February 22, 1864. A part of the report is to the following effect: "But the whole evidence showing that the lessors have repeatedly failed to deliver brine to the state furnaces, according to the terms of their contract of lease, whereby said furnaces could not be worked up to their full boiling capacity, and were in fact often 'blown out' entirely for the want of a supply of salt water, the committee are clearly of opinion that there is a legal liability resting on Stuart, Buchanan Co., to make good to the state the loss resulting from this failure on their part to comply with their contract. The tenor of the evidence estimates the loss from this source at fifteen thousand to twenty thousand bushels of salt." "Report of the Joint Committee on Salt Relative to the Contract with Stuart, Buchanan Company", *Virginia Legislative Documents,* 1863-64, Document No. 26, p. 3.

For statement with regard to the conditions laid down by the lessors for renewal of the lease, see *Ibid.,* pp. 4-5; see also "Answer of Superintendent J. N. Clarkson Relative to the Condition of the Salt Works," Document No. 29, p. 3. Document No. 16, sometimes bound with the *Virginia Senate Journal,* Regular Session, 1863-64, is the same as Document No. 26.

29. *Session Laws of Virginia,* 1863-64, chapter 6. It is interesting to note early conservation features. Section 4 of the act declared that the superintendent might supply wood, but must leave at least one-fifth of the timber standing on any tract of average value.

30. Governor Letcher states this view in his annual message to the legislature, December, 1864, *"Documents Relative to the Subject of Salt,"* Document No. 43, p. 10.

31. Note that General Pemberton was obliged to seize the New Orleans, Jackson, and Great Northern Railroad with all its rolling stock. See letter of the president of the road to Governor Pettus, December 1, 1862, Pettus Papers, volume 58, Mississippi Archives.

32. *Ibid.,* Letter of the Vice President of the Southern Railroad to Governor Pettus, January 14, 1863, *ibid.,* series E, volume 59.

33. An official of the Southern Railroad begged Governor Pettus to help supply more rolling stock. An interview with General Smith encouraged the railroad official to promise the governor that no further detention in the transportation of salt and wood would occur. Found in a postscript to a letter to Governor Pettus, (No date). But still the salt did not move.

34. President of the Southern Railroad to Governor Pettus, December 2, 1862; member of the Cobb Manly Company to Rives, December 8, 1862, *ibid.;* Governor Pettus's report to the Mississippi Assembly, December 20, 1862, *O.W.R.,* series IV, volume II, pp. 250-51; and letter of the vice-president of the same road to Governor Pettus January 14, 1863, Pettus Papers, series E, volume 59, Mississippi Archives.

For the soldier's outburst quoted above, see W. H. Hardy to Governor Pettus, December 2, 1862, *ibid.,* volume 58.

35. There is evidence as early as May 3, 1862, that the movement of troops at Corinth was preventing the shipment of salt. See letter of General D. C. Green to Governor Shorter from Mobile, May 3, 1862, Executive Correspondence, Alabama Archives.

A rather interesting story is revealed in a letter from Governor Shorter to Quarter-Master General Green, dated May 31, 1862, showing how a certain Judge Walker, who was connected with the railroad in some capacity, succeeded in bringing through to Mobile salt after the state agent had failed. *Ibid.*

36. Letter of N. Kile to Governor Shorter, February 13, 1863. *Ibid.;* letter of G. R. Clopper to John Graham, August 17, 1863, *ibid.,* series W.

37. See the letters of General R. E. Lee to J. B. Floyd of September 4 and 9, 1861, *O.W.R.*, series I, volume LI, part 2, p. 270; and Floyd's reply of September 9, *ibid.*, pp. 286-87.

38. Colonel W. E. Peters so reported to Adjutant-General Cooper, *ibid.*, pp. 463-64.

39. *Ibid.*, volume XV, p. 860. General Taylor erected a defense work on the Atchafalaya River for this purpose at the Butte à-la-Rose, where the numerous branches of that stream unite in one channel. *Ibid.*, p. 873. The chief of subsistence was doubtful, already in January, whether the line could be kept open, *ibid.*, volume XXIV, part 3, p. 591.

40. Pollard, *The Lost Cause*, p. 483.

41. *O.W.R.*, series IV, volume II, p. 175.

42. See his message to the Georgia Assembly of November 6, 1862. *Georgia Senate Journal*, Regular Session, 1862, pp. 30, 31. He also complained that his protest to Governor Letcher, written as soon as he saw the act published in the newspapers, had received no reply. A reply, however, was on its way. *Ibid.*, pp. 63-64. Echoes were heard in the confederate congress.

43. *O.W.R.*, series I, volume LII, part 2, p. 384. The letter is dated October 31, 1862.

44. Woodfin to ex-Governor Swain, November 17, 1862. Walter Clark Manuscript, volume IV, North Carolina Historical Commission.

45. The appeal took the form of a joint resolution, *Session Laws of Georgia*, Regular Session, 1862, No. 3. See also a letter of E. W. Chastain to Governor Brown, March 5, 1863, letter in the Georgia Archives.

46. *O.W.R.*, series I, volume LII, part 2, p. 385.

47. *Ibid.*

48. *The Richmond Dispatch*, October 2, 1861. The same paper quoted a rumor that powerful combinations of capitalists prevented the transmission of salt over the railroads.

49. Jones, *A Rebel War Clerk's Diary*, volume II, p. 401; Daniel Worth to President Cowan of the W. C. and R. Railroad, February 9, 1864, Executive Papers, Z. B. Vance, North Carolina Historical Commission.

50. See letter of the salt agent of Cherokee County, Murphy, to Governor Vance, April 14, 1863, *Ibid.* Note also the following almost pathetic plea from G. N. Bristol, agent for Clay County to Governor Vance, dated March 10, 1863: "I have Bin two trips to Saltville, Va. I have Got the salt from Mr. Woodfin But it seems I cant Git it shiped & consequently we have to do without. The Dificulty they say is to Git it shiped from Saltville to Bristol. Mr. dodomede the superintendent on that Part of the Road utterly refuses to ship our salt cherokee is in the same fix my object in writing you is to implore your aid Perhaps you could do or say something that would cause him to ship our salt & relieve a long suffering community now suffering for salt. a Plenty at the works." Vance indorsed on the letter, "It is utterly impossible for the Govr. to get this salt shipped, as the same difficulty exists in this part of the state." *Ibid.*

51. Governor Vance to Colonel A. C. Myers, December 26, 1862, Vance Letter Book, 1862-63, p. 75, North Carolina Historical Commission. Of

course, these states had already embarked on the method of hauling salt by wagon for long distances as detailed above in chapter VII.

52. Woodfin to Governor Vance, April 4, 1863, Executive Papers, Z. B. Vance, North Carolina Historical Commission.

53. No later than August 26, 1862, Governor Brown was thinking of the possibility of sending a train for this purpose, though he did not wish to do so, as the trains were needed at home. See letter of Governor Brown to B. H. Bigham, president of the Alabama Salt Manufacturing Company, Letter Books-Governors', 1861-65, p. 336, Georgia Archives.

54. It is apparent that Governor Shorter had appealed to Governor Brown for aid from the Georgia State Railroad in transporting Alabama salt from Virginia, because a reply from the latter under date of November 6, 1862, expresses his willingness to aid Alabama after the Georgia salt had been brought out, provided he could recover the railroad stock from General Bragg. *Ibid.,* p. 354.

55. Though the joint resolution authorizing such action did not become legally effective until November 25, when it was formally approved, the governor became active in the matter on November 22. *Session Laws of Georgia,* Regular Session, 1862. Joint Resolution, Nos., 11 and 14. The state spent $50,000 to send the trains to Saltville.

56. Letter Books-Governors', 1861-65 pp. 365, 366, 367, 383, Georgia Archives. Some of the roads as a return favor asked the right to buy and bring out salt for their own use, (*Ibid.,* pp. 365, 366, 388,) which request was granted in, at least, one instance. *Ibid.,* p. 388.

57. *Session Laws of Georgia,* Regular Session, 1862, Joint Resolution, No. 21, signed December 4, 1862. By April, 1863, 40,000 bushels of salt, made for Georgians, were awaiting transportation, according to the Joint Committee on Salt Supply, *Georgia Senate Journal,* Extra Session, 1863, p. 127.

58. See Governor Brown's Annual Message of November 6, 1862, *Georgia Senate Journal,* Regular Session, 1862, p. 30; and *The Confederate Records of the State of Georgia,* volume II, p. 230. To obtain the privilege of free transportation over the state roads, citizens were merely required to file affidavit that the salt was for home consumption only, Letter Books-Governors', 1861-65, p. 383. Georgia Archives. Governor Brown to Colonel Jno. Harris, Nov. 27, 1862.

There is a slight discrepancy between the directions given to Temple and Bigham on May 23, 1863, and those given to Major I. S. Rowland, of the state railroad on April 23, 1863, on the same subject, *ibid.,* pp. 477, 467. I have followed those of the later date.

59. Governor Brown to Bramer, May 27, 1863, *Ibid.,* p. 484, Georgia Archives.

60. J. M. Quail to Governor Vance, November 24, 1863, Executive Papers, Z. B. Vance, North Carolina Historical Commission.

61. "Documents Relative to the Subject of Salt," Document No. 43, pp. 12, 14.

62. Letter Books-Governors', 1861-65, p. 388, Georgia Archives. Governor Brown was careful to stipulate that the salt be loaned the lessees by weight and not in sacks, as it might be difficult to secure the return of the sacks. *Ibid.,* pp. 403, 417.

63. See the report of the Joint Committee on Transportation of the

Georgia Assembly, *Georgia Senate Journal,* Extra Session, (March) 1863, p. 121. Brown sought to secure from the war department permits for the free passage of such a train. Letter Books-Governors', 1861-65, p. 365, Georgia Archives.

64. See the letter of Superintendent Dodomead to Major B. H. Bigham, March 24, 1863, *Georgia House Journal,* Extra (March) Session, 1863, p. 113.

65. See letter of President Wallace of the Wilmington and Weldon Railroad to Governor Vance (no date). Executive Papers, Z. B. Vance, North Carolina Historical Commission; also the report of the Board of Public Works of Virginia, "Report of the Joint Committee in Respect to Salt for Georgia," *Virginia Legislative Documents,* Extra Session, 1863, Document No. 10, p. 5.

66. See letter of Woodfin to Governor Vance, April 8, 1863, Executive Papers, Z. B. Vance. North Carolina Historical Commission.

67. Governor Brown in presenting the proposition to the assembly at an extra session in March-April, 1863, asked a further authorization to purchase or impress trains from some of the private companies of the state, as he held the resolutions to that end of the preceding session were merely temporary in their nature. Message of April 3, 1863, *Georgia House Journal,* Extra Session, March-April, 1863, pp. 102-03. The pay offered for the hire of the train was ten dollars a day for the engine and two cents a mile run by each car, while Georgia was to be responsible for all major repairs.

68. That order was as follows:

(1) Salt manufactured by M. S. Temple for the state of Georgia and wood and supplies for his works.

(2) Planters Salt Manufacturing Company.

(3) Georgia Salt Manufacturing Company.

The last two to be served in equal measure.

(4) All other persons making or buying salt for consumption in Georgia.

See letter of Governor Brown written jointly to Colonel B. H. Bigham and M. S. Temple under date of May 23, 1863, Letter Books-Governors', 1861-65, p. 477, Georgia Archives. These two men were to make the best arrangements they could with the superintendent of the Virginia and Tennessee Road as to payment for the use of the train.

69. See W. W. Walley's circular letter to the governors, dated April 8, 1863, Executive Papers, Z. B. Vance, North Carolina Historical Commission; also letter of Governor Brown to John P. King, president of the Georgia Railroad, of April 20, 1863, Letter Books-Governors', 1861-65, p. 464, Georgia Archives; Governor Shorter to Secretary Seddon, April 30, 1863, Executive Correspondence, Alabama Archives.

70. Governor Letcher to Governor Vance, December 10, 1862, Executive Papers, Z. B. Vance, North Carolina Historical Commission.

Evidently Georgia did not accept the method of retaliation suggested by one irate citizen of that state, for we find Governor Brown promising Governor Smith of Virginia on November 7, 1864, to do his best to spare a train occasionally to carry provisions to Virginia, if the former would bring Georgia salt to Richmond. Letter Books-Governors', 1861-65, pp. 818-19, Georgia Archives.

71. Governor Letcher to Governor Vance, December 10, 1862, Executive Papers, Z. B. Vance, North Carolina Historical Commission.

72. J. M. Quail to Governor Vance, November 24, 1863, *Ibid.*

73. See the "Report of the Joint Committee on Salt Relative to the Difficulties about Foreign Salt Trains etc.", inclosed with letter of Governor Smith to Governor Vance, March 9, 1865, Z. B. Vance Papers, volume VII, North Carolina Historical Commission. It is also printed in "Documents Relative to the Subject of Salt," Document No. 43 among the Virginia legislative documents.

74. Governor Brown's commission to Bigham is dated September 23, 1863, Letter Books-Governors', 1861-65, p. 548, Georgia Archives. Governor Brown seems also to have commissioned A. K. Seago on October 11, 1863, to manage the shipment of salt and to have the engine, *Texas,* repaired, *Ibid.,* p. 549.

75. *O.W.R.,* series I, volume LI, part 2, pp. 1058-59; *Session Laws of Virginia,* Called Session September-October, 1863, Joint Resolution No. 7.

76. Up until the time that Stuart, Buchanan and Company secured such a monopoly of the railroad, transportation of Virginia salt had kept nearly even pace with production. "Biennial Report of the Board of Public Works", *Virginia Legislative Documents,* Document No. 15, pp. 6-7; *Session Laws of Virginia,* Called Session, 1863, Joint Resolution, No. 6.

77. See letter of Hawkins, a North Carolina agent at Saltville, to Governor Vance, November 30, 1864. Executive Papers, Z. B. Vance, North Carolina Historical Commission. It might be noted that if the demand to have arrearages made up had been enforced, the Georgia train would have been kept busy a solid month without carrying a pound for Georgia. "Documents Relative to the Subject of Salt," Document No. 43, p. 72.

For the best explanation of this entire question of the controversy between Virginia and the other states, especially North Carolina, see *Ibid.*

78. Governor Smith regarded the language, with considerable justice, it must be admitted, as so offensive that he did not transmit the protest to the Virginia legislature as likely to "foment interstate ill-will." The protest was signed by L. M. McClung for Tennessee and Alabama; by P. B. Hawkins for North Carolina; by L. D. Palmer and B. H. Bigham for Georgia. It is to be found in Executive Papers, Z. B. Vance, North Carolina Historical Commission, but has not been printed among the Virginia documents. Governor Smith evidently suppressed it from the Virginia records.

79. All the trains had left at these interferences. There was some loss to Virginia in the policy pursued, for the visiting trains had helped in sudden emergencies with the moving of troops, while the movement of salt had kept the price down to some extent. It is interesting to know that there had been ten visiting trains in all running on the Virginia and Tennessee road, some of which had been on the road for two years. They were now driven to hire themselves to other Virginia roads. Based on the statement of R. L. Owen, president of the Virginia and Tennessee Railroad Company, made December 1, 1864. "Documents Relative to the Subject of Salt." Document No. 43, p. 82.

80. *Public Laws of North Carolina,* 1864-65, Joint Resolution, pp. 74-75.

81. *O.W.R.*, series I, volume LI, part 2, p. 1057. The restrictions, he remarked caustically, would be removed as soon as Virginia revoked her unprecedented orders. *Ibid.*

82. *Ibid.*, pp. 1058-60. Governor Smith was sure that North Carolina was better provided with salt than Virginia. Governor Smith to Governor Vance, January 28, 1865, *ibid.*, p. 1060.

83. See governor's annual message, December, 1864, *Governor's Message and Accompanying Documents of Virginia,* 1864, Document No. 1, p. 20.

84. *Ibid.*

85. The governor of Virginia had previously recommended in his message of 1864, an act to prohibit the superintendent from any interest in the sale of salt in order to bring consistency with the law which prohibited any interest on that official's part in the making of salt and to protect him against suspicion of appropriating transportation for himself, *Ibid.*, pp. 19-20. The executive reached this position, despite the fact that the attorney general ruled that the superintendent was not violating the law.

86. Z. B. Vance Papers, North Carolina Historical Commission.

87. "Documents Relative to the Subject of Salt", Document No. 43, p. 12.

88. Based on the final report of Woodfin, March 21, 1865, Executive Papers, Z. B. Vance, North Carolina Historical Commission.

## CHAPTER X

1. *Official War Records,* series I, volume XVII, part 2, p. 141.

2. *Ibid.*, series III, volume II, p. 402. He ordered a captain tried for aiding the enemy by furnishing them salt, "a contraband article," *Ibid.*, series I, volume XVII, part 2, p. 178.

Orders which were issued at Nashville on May 29 preceding imply, but not so explicitly, that salt was contraband. They declare that since it was reported that salt, bacon, coffee, iron, medicine, and other goods were being sold in the city and so finding their way to the enemy, no goods might be taken from the vicinity of Nashville toward the enemy's lines without a permit of the provost marshal of that city. *American Annual Cyclopedia,* 1862, pp. 597-98.

3. *O.W.R.*, series III, volume II, p. 402.

4. He felt that such a list would save officers a world of trouble, *Ibid.*, series I, volume XXXI, part 1, p. 736.

5. *Ibid.*, volume XV, p. 899. The order declaring salt contraband and ordering salt and salt works destroyed wherever found was attributed to President Lincoln, but there is no evidence that he even knew of it until it was promulgated. Andrews points out that such an order was more in accord with Stanton's character. *Women of the South in War Times,* p. 28.

6. Many general statements are to be found naturally. Bradlee, for instance, says that a great deal of salt was smuggled in from the north in exchange for cotton, *Blockade Running during the Civil War,* p. 47.

7. Howe, *Sherman's Home Letters,* p. 229. Cox's suspicion, as early as

September, 1861, of a secret trade in salt and beef in western Virginia between Kanawha and Wytheville, which prompted a project to sweep the whole territory, was probably well-founded, but cannot, of course, be asserted as a fact, *O.W.R.*, series I, volume LI, part I, p. 484. In the same category must go Colonel Packham's belief of a large salt business by the rebels in Rock County, Missouri, about the same time, *Ibid.*, volume VII, pp. 6, 472.

8. *Ibid.*, volume XVI, part 1, p. 861.

9. *Ibid.*, volume XVII, part 2, p. 15.

10. *Ibid.*, pp. 141, 187.

11. *Ibid.*, series III, volume II, p. 349. In Sherman's usual pointed style he remarks that if the avenues of trade should be opened, Memphis would be better to the enemy under union control than before it was taken.

12. *Ibid.*, series I, volume XVII, part 2, p. 151.

13. *Ibid.*, volume XX, part 2, p. 172.

14. *Ibid.*, volume XXX, part 3, p. 9.

15. *Ibid.*, volume XVIII, p. 146.

16. *Ibid.*, volume XLI, part I, p. 884.

17. *Ibid.*, series IV, volume II, p. 57.

18. *Ibid.* It must be added that while common sense indicates that this very type of trade was undoubtedly conducted, the writer has found no further evidence.

19. *Ibid.*, series I, volume XXXI, part 3, pp. 833-34.

20. A generous officer furnished her money to enable her to return to her home, *Ibid.*

21. For an illustration see *Ibid.*, volume XLVIII, part I, p. 1423.

22. Series G, No. 201, Mississippi Archives.

23. For such illustrations see *O.W.R.*, series I, volume XXIV, part 3, pp. 177-78; series IV, volume II, p. 854 ffl.

Two men offered to make a gift of 2,000 sacks of salt to the families of poor soldiers in return for the privilege of shipping out 1,000 bales of cotton. Thomas G. Davidson to Governor Pettus, December 20, 1862, Pettus Papers, volume 58, Mississippi Archives.

24. Governor Clark to Major Irehill, July 15, 1862, Gov. Clark, Letter Books, 1861-62, p. 372, North Carolina Historical Commission.

25. Aide to Governor Vance to Secretary Randolph, October 15, 1862, Vance Letter Book, 1862-63, September-October, pp. 27-28, North Carolina Historical Commission.

26. W. J. Gillam to Governor Vance, November 11, 1862, Executive Papers, Z. B. Vance, 1862, North Carolina Historical Commission.

27. General S. C. French to Governor Vance, February 5, 1863, Vance Letter Book, 1862-63, September-October, p. 129, North Carolina Historical Commission.

28. Summers, *History of Southwest Virginia,* p. 533.

29. Governor's Letter Books, 1861-65, p. 747, Georgia Archives.

30. *Ibid.*, p. 751.

31. A communication from the confederate collector at Tangipahoa to Governor Pettus, dated December 3, 1862, seems to indicate that he so interpreted his orders. Pettus Papers, volume 58, Mississippi Archives.

32. Bradlee, *Blockade Running,* p. 47; *O.W.R.,* series IV, volume I, p. 237.

33. Taylor, *Running the Blockade,* p. 18; *The South in the Building of the Nation,* volume V, p. 668.

34. September 26, 1861.

35. *The Western Democrat* (Charlotte, North Carolina), November 26, 1861.

36. October 2, 1862; *The Daily Mississippian* (Jackson, Mississippi). October 8, 1862.

37. The *Clarke County Democrat.* (Grove Hill, Alabama) August 14, 1862.

38. *The Daily Southern Crisis* (Jackson, Mississippi), March 24, 1863. The mere fact that such an insignificant amount was noted speaks eloquently of the scarcity of the commodity. Its grains were counted almost like medicine. Other instances are noted in *O.W.R.,* ser. I, volume LI, part 2, pp. 63-65 and Bonham, "British Consuls in the Confederacy," *University of Columbia Studies in History, Economics, and Public Law,* volume XLIII, p. 56.

39. Governor Shorter had learned from a military officer that the army did not need the salt and so he tried to secure it for the state of Alabama, Governor Shorter to Green, Quarter-Master General, July 28, 1862, Executive Correspondence, Alabama Archives.

40. The agent for the schooner was greatly concerned for fear that the embargo recently laid by the state might prevent the salt from leaving North Carolina, Unsigned letter to Governor Vance of November 14, 1862. Executive Papers, Z. B. Vance, September-November, 1862, North Carolina Historical Commission. In March, 1863, a schooner bound for Savannah had 310 sacks of salt aboard, *O.W.R.,* series I, volume LIII, p. 87.

Merchants of a port like Wilmington tried to move on their imported supplies of salt to points where there was no competition with a domestic supply. Hence we find a citizen reporting to Governor Vance that several weeks previous to the date of writing there were in Wilmington 50,000 to 60,000 bushels of imported salt and that efforts were being made to send most of it to Augusta, Georgia, where prices were higher, because there was no manufacturing of salt there. J. W. Sinclair to Governor Vance, October 27, 1862, *Ibid.*

41. *O.W.R.,* series I, volume XVIII, 780-81; Bonham, "British Consuls in the Confederacy," *University of Columbia Studies,* volume 43, pp. 115-116. They were delivered to the British consul and arrangements were made with Fraser and Company to give them employment on one of that company's boats in order to enable them to reach home.

42. *O.W.R.,* series I, volume XXIX, part 2, p. 582. Note also the capture of the expedition, *Ibid.,* volume LIII, p. 87.

One interesting effort to bring in Liverpool salt via Halifax died "aborning." An investment was made in November, 1862, to purchase vessels in order to import Turk's Island salt. It would appear to have been made by the confederate government, for the sums amounting to $47,459, are duly recorded by J. L. Locke, major in command. He adds at the foot of the paper the following significant words: "The above enterprise failed in consequence of the inability of Agent and Captains to proceed to Halifax by Steamer, Hem." Unpublished manuscript among the *Official War Records* of the War Department of the United States. The *Milan* from Liverpool with 1,500 sacks of salt was also lost. *The American Annual Cyclopedia,* 1861, p. 586.

43. This supply seemed to enter Texas near Presidio del Norte, *O.W.R.,* series I, volume L, part 2, p. 377. Amounts as large as 3,000 bushels of salt entering Houston from Mexico were noted. See *The Clarke County Democrat* (Grove Hill, Alabama) December 5, 1861.

44. Governor Pettus had already paved the way with the central government by several communications in which he begged that no decision be made until all the facts had been presented. He wrote President Davis on October 8 and again on October 17, *O.W.R.,* series I, volume LII, part 2, pp. 372, 412; series IV, volume II, p. 126.

45. Pettus frankly stated that General Butler, then in command of the federal forces at New Orleans, might have been bribed by the Frenchman, but he did not consider that his affair. When he widened the scope of the negotiations to include any responsible English house, he still insisted that there must be assurance over the consular seal that the cotton would be shipped to some port in Europe. Governor Pettus to C. Steel, November 25, 1862, Pettus Papers, volume 58, Mississippi Archives.

46. He thought that twenty sacks of salt to one bale of cotton should have been arranged, since salt was bringing only one dollar a sack in New Orleans whereas cotton commanded $160 a bale. The president did not consult his secretary of war on the matter. We are indebted for these intimate details to Jones, the war clerk. See Jones, *A Rebel War Clerk's Diary,* volume I, p. 185; see also Cooper to Governor Pettus, December 1, 1862, Pettus Papers, volume 58, Mississippi Archives.

47. Contracts were made with John C. Burdon and George Leary, British subjects, and with Adolphus Mennett, a Frenchman. For the contracts see Pettus Papers, volume 58, Mississippi Archives; *O.W.R.,* series I, volume XV, p. 617. The former were dated December 6, and the one with Mennett November 27, 1862.

48. A perfect ripple of excitement went out through the state while these contracts were under consideration. Either the state government expected to buy up cotton for the exchange on promises to deliver the salt later, or the people, hearing rumors of the negotiations, assumed that exchanges with the citizens were to be made immediately, for the governor was bombarded with a barrage of letters of inquiry concerning the exchange of cotton for salt and of bacon for salt. See the file of correspondence on the subject in Pettus Papers, volume 58, Mississippi Archives.

49. For Governor Pettus's own account of the transaction see his message to the assembly, *O.W.R.,* series IV, volume II, p. 923.

The fifty bales of cotton bought to pay Mennett were finally turned over to Dr. Luke Blackburn, one of the medical commissioners of the state, to be exchanged at Havana for medical supplies. Rowland, *Mississippi,* II, pp. 591-92. It is not regarded as necessary to go into the proposition of C. A. Barrière and Brother of New Orleans to furnish the confederate government with 100,000 sacks of salt on condition that with the proceeds of the salt they might buy cotton on the Mississippi. This transaction is set forth in *O.W.R.* series IV, volume II, pp. 173-74; Jones, *A Rebel War Clerk's Diary,* volume I, p. 187.

It is interesting that the Richmond government found itself virtually forced to grant the same privilege to the customs collector of Louisiana, *O.W.R.*, series IV, volume II, p. 242.

## CHAPTER XI

1. A reconnoitering expedition sent out by General Burnside on February 19, 1862, reported that there were no organized salt establishments on the sound, only a few old iron kettles being found in operation to obtain a supply for local use. The inhabitants were greatly alarmed, but were reassured by the federals that there was no purpose to harm them. *O.W.R.*, series I, volume IX, p. 366. As elsewhere, the salt works were rebuilt at the same site, Wales Head, only to be destroyed again in February, 1863. They were producing about fifty bushels a day at that time, *Ibid.*, volume XVIII, p. 148.

2. *Ibid.*, pp. 415 and 416. General Foster reported in October, 1862, from Newbern, North Carolina, "I have destroyed all the salt-works within reach of me.", *Ibid.*, p. 416.

3. See above, p. 101. Governor Clark of North Carolina had written to President Davis as early as March 17, 1862, expressing the conviction that the next move of General Burnside would be to destroy the last connection of North Carolina with the ocean, *Ibid.*, volume IX, p. 446. Yet the blow did not fall for more than two years.

4. *The Correspondence of Jonathan Worth,* volume I, p. 309; Daniel Worth to Governor Vance, *North Carolina Legislative Documents,* 1863-64, Document No. 10, p. 2.

5. The devices resorted to in order to locate salt works are interesting. A certain adventurer from Marianna, Florida, was suspected of having salt wells. Hence, federals fell into the habit of landing for predatory incursions. One day a party asked this man what his business was and wished to see his establishment. The owner put them off and thus saved his wells for a time. Hopley, *Life in the South,* volume II, pp. 309-10.

6. A small raid by Lieutenant-Commander Bradish from the vessels, *Penguin* and *Henry Andrew,* on March 22, 1862, against some small salt-works near New Smyrna seems a detached event, which apparently did not affect the Florida activity in the least.

7. *Naval War Records,* series I, volume 17, pp. 310, 311, and 316; *New York Herald*, October 30, 1862. They had a capacity respectively of 200 and 216 bushels.

8. *Naval War Records,* series I, volume 17, p. 318.

9. For the whole episode see *Ibid.*, pp. 316-19.

10. *Ibid.*, volume 19, p. 373. Based on the statement of Lieutenant-Commander Hart.

11. *Ibid.*, p. 374.

12. *Ibid.*, pp. 374-75. For the full report of Lieutenant-Commander Hart see *Ibid.*, pp. 373-76.

13. *Ibid.*, p. 377. It is of passing interest that the men at this point had heard of the attacks by Admiral Lardner's squadron the preceding Sep-

tember and that some of the salt workers had moved up here from salt works further south on the coast after the raid of October 4 on Cedar Keys. *Ibid.,* p. 375.

14. *Ibid.* The unionists chased the escaping salt-makers for three miles across a swamp and overtook them, breaking up the kettles in their wagons.

15. *Ibid.,* p. 375. Hart found himself unable to destroy all the works found because of the exhaustion of his men. He thus failed to visit California Inlet, where there were said to be 1,000 salt-makers under the protection of a large body of cavalry, or those on the east end of Rosa Island, which they passed on the return journey to Pensacola, *Ibid.*

16. *Ibid.,* p. 377.

17. Commander Hart gives the total capacity of the salt works destroyed as 21,640 gallons a day. *Ibid.,* p. 379.

18. *Ibid.,* volume 17, pp. 467-72. This raid cost the confederates the loss of 200 bushels of salt which were scattered on the shore. The Comte de Paris notes also a raid in July of this year on the works on Marsh Island, near the Ocklockonnee River in Apalachee Bay, *History of the Civil War in America,* volume IV, p. 389.

19. *O.N.R.,* volume 17, pp. 593-94, 596.

20. The government had here 27 buildings, 22 large boilers, and 300 kettles, each averaging 200 gallons; 2,000 bushels of salt, and store-houses containing three months' provisions, *Ibid.,* p. 597.

21. *Ibid.*

22. *Ibid.,* pp. 594, 598.

23. The writer has arrived at this figure by adding up the capacity at the various works noted in the reports and then reducing gallon capacity to bushels on the basis of the ratio for salinity of the water from St. Andrew's Bay, which has, fortunately, been recorded.

24. *Ibid.,* p. 598.

25. *Ibid.* This statement of the seriousness of the blow was quoted by the *New York Herald,* January 5, 1864.

26. This is not the same Browne alluded to earlier. The name of the former was George W., while the one here referred to signed his reports W.R.

27. For a full account of the destruction wrought from December 11 to 18, see *Official Naval Records,* series I, volume 17, pp. 593-601; Moore, *Rebellion Record,* volume 8, p. 280; *New York Herald,* January 7 and 19, 1864. As is not strange, the estimates of the loss differ. A paper inclosed with Browne's report of December 11 gives far larger figures than he gives in the text of his report. Compare also *O.N.R.,* volume 17, p. 598 and p. 600. The *New York Herald* seems to improve even on Browne's most generous figures. As usual, the writer adopts the more conservative figures as the safer.

The delighted comments of the *New York Herald* of January 5 may be noted: "Salt works are as plentiful in Florida as blackbirds in a rice field."

28. See the report of Browne, *O.N.R.,* series I, volume 17, pp. 646, 647.

29. Browne states as his belief that the boilers and kettles alone on West Bay must have cost $146,883. *Ibid.,* p. 646.

30. See *Ibid.,* pp. 646-47. The date of the first raid can be established

from the abstract of the log of the *Restless,* and is as given above, *Ibid.,* p. 648.

The productive capacity was 19,350 gallons of salt or 623 bushels, translating wet into dry measure.

31. February 17 and 27 were the exact dates.

32. *Ibid.,* p. 648.

33. *Ibid.,* p. 649.

34. *Ibid.,* pp. 676-678.

35. *Ibid.,* p. 683. In a later report he repeated a rumor that the confederate government intended to erect works on the east arm of St. Andrew's Bay.

36. The ensign on this occasion, May 24, 1864, destroyed eleven small works of about sixty kettles in all, *Ibid.,* p. 707.

37. *Ibid.,* p. 719.

38. *Ibid.* See also *New York Herald,* November 8, 1864. It must be added for accuracy that the *Restless* had changed commander by the last date.

39. *O.N.R.,* series I, volume 17, p. 811; Report of the Secretary of War, 1865-66, p. 351.

40. *New York Herald,* December 17, 1864.

41. *Ibid.,* September 20, 1864.

42. Shorter to Randolph, September 22, 1862, Executive Correspondence, Alabama Archives. On December 15, in an effort to arouse the government to a sense of its own dangers, he commented to the secretary of war on the fact that Mr. Clendenin's works had been broken up by the foe. Clendenin was superintendent of the government works at St. Andrew's Bay at the time. *O.W.R.,* series I, volume XIV, p. 716.

43. *Florida Session Laws,* Session of November-December, 1862, Joint Resolution, No. 30.

44. *O.W.R.,* series I, volume XIV, pp. 753-4, statement of W. Fisher, the designated officer.

45. On the section of coast between the Apalachicola and Choctawhatchee Rivers he found in part an almost desert country of nearly fifty miles. He did not apprehend a raid in that quarter beyond St. Andrew's Bay. *Ibid.,* pp. 730-731.

46. Shorter was pessimistic about the civilian population. He feared that under the inducements which would be held out, a large portion would save their little property by giving their adherence to the enemy. Shorter to General Buckner, December 31, 1862. Executive Correspondence, Alabama Archives.

47. *O.W.R.,* series I, volume XV, pp. 947-8. Shorter had a special interest in staying the enemy in the coast region of Florida, for federals had already made a raid up the Choctawhatchee River into Coffee County, Alabama. The south-eastern part of the state was also peculiarly liable to hostile incursions from Pensacola. See letter of January 10, 1863, from Shorter to Davis, *Ibid.,* p. 939.

48. Governors' Letter Books, 1861-65, p. 525. Georgia Archives. Governor Brown represented to Secretary Seddon on August 10, 1863, that he was being importuned by a large number of slave owners who were making salt on the Georgia coast to protect their works and slaves against

the constant threats of federal raiders. Governors' Letter Books, 1861-65, p. 525, Georgia Archives.

49. View of a citizen resident in Tallahassee. *O.W.R.*, series I, volume XIV, p. 753.

50. Governors' Letter Books, 1861-65, p. 352, Georgia Archives.

51. There is evidently some error of geography in the record. The report of the officer locates Bluffton on the May River in Georgia, though it is obviously in South Carolina, on a deep inlet south of the May. *O.W.R.*, series I, volume XIV, p. 125.

52. *Ibid.*, p. 124-126. As an evidence of the destructiveness of war, it might be noted that the federals carried off a large quantity of furniture from the deserted houses at Bluffton.

53. *Ibid.*, p. 190.

54. See report of Brigadier-General Saxton of November 12, 1862. *Ibid.*, pp. 189-90; and of Quarter-Master Meigs to the secretary of war of November 18, 1862, *Ibid.*, series III, volume II, p. 808.

55. *Ibid.*, series I, volume XIV, pp. 197, 315. The latter was unsuccessful.

56. Moore, *Rebellion Record,* VII, p. 311.

57. *O.N.R.*, series I, volume 19, pp. 380-382.

## CHAPTER XII

1. *Louisville Daily Journal,* January 4, 1862.

2. The writer is accepting the figures of the officer who directed the work of destruction rather than that of the commanding general. Compare *O.W.R.*, series I, volume XVI, p. 1153 and p. 1150.

3. A story was related to General W. S. Smith in connection with the destruction of the salt works, which indicates a passionate devotion to the union on the part of some of the citizens of Kentucky, even when they were the victims of the war measures. The noble conduct of some of those interested in the works, especially of Mrs. Garrard, who expressed her entire willingness that "not only the valuable property, but all else that she and her husband (a colonel in our service) owned, might be destroyed if such destruction would help to restore the Union," constrained an earnest recommendation that prompt restitution should be made for the damage done. *Ibid.*, p. 1149.

4. *Ibid.*, volume XX, part 2, p. 395. It is, of course, possible that broken communications had prevented his knowing by November 8 of the raid of October 23-24, but the rest of the letter shows the colonel well informed on military movements. Furthermore, the officer reporting on the Goose Creek raid states that the time allowed was too limited "to thoroughly complete the job." *Ibid.*, volume XVI, part I, p. 1150.

5. Hopley, *Life in the South by a Blockaded British Subject,* volume I, p. 303-04. A southern argument held that these wells made a million and a half bushels of salt a year. Among the pork packers of Cincinnati the Kanawha salt was preferred to any other. *Ibid.*, p. 365.

6. For the campaign of 1861 see Shotwell, *The Civil War in America,* volume I, pp. 112-16.

7. For Lee's congratulatory letter to Loring see *O.W.R.*, series I, volume XIX, part 2, pp. 625-26.

8. Pressure was put on Secretary Randolph to attempt the recovery of the Kanawha salt works. Note the letter of the former attorney general of the Central Railroad, *Ibid.*, volume XIII, part 3, p. 937.

9. Now known as Pearisburg, Va.

10. Only one of the furnaces was burned. *Ibid.*, volume XIX, part I, pp. 1083-84.

11. The confederates captured 700 barrels of salt. *Ibid.*, volume LI, part 2, p. 618.

12. See the report of General Loring, dated September 14, *Ibid.*, volume XIX, part I, p. 1071.

13. For the official reports on the entire campaign see *Ibid.*, pp. 1057-90.

14. *Ibid.*, volume LI, part 2, p. 621.

15. Pierpont was governor, it will be recalled, of the fragment of Virginia which remained loyal to the union.

16. *Ibid.*, volume XIX, part 2, p. 174. The date is not clear in the record, but it must have been on October 3.

17. General Loring wrote Secretary Randolph October 7 as follows: "As I have before written to you, this valley could only be held until high waters gave access to gunboats and easy transportation to the enemy . . ." *Ibid.*, p. 656.

18. *Ibid.*, Quoted by General Loring to Secretary Randolph.

19. See Randolph to Echols, same date. *Ibid.*, p. 666.

20. *Ibid.*, volume XX, part 2, pp. 384, 293.

21. Jones, *A Rebel War Clerk's Diary*, volume I, pp. 185-86.

22. *O.W.R.*, series I, volume XXV, part I, pp. 90-106. Other threats were made by the confederates more directly at the Kanawha Valley, one by General Sam Jones on the lower Kanawha in March, 1863, *Ibid.*, volume XXV, part 2, p. 668; and one by Nounnan, September 23-October 1, 1864, *Ibid.*, volume XLIII, part I, pp. 641-44. For the fear engendered among the citizenry, see letter of a judge to the governor, dated October 25, *Ibid.*, part 2, pp. 490-91.

23. *Ibid.*, volume XV, p. 855.

24. *Ibid.*, p. 175. Report of General Taylor, dated November 9, 1862.

25. It seems passing strange that the discovery of the first salt mine in the confederacy created so little furor among federals that General Butler's manoeuvres were predicated on the assumption that the salt works at New Iberia were of the ordinary evaporating variety. See the comments of *The Daily Southern Crisis* (Jackson, Mississippi), January 24, 1863. Southern papers naturally rejoiced at his discomfiture over his failure.

26. The date given by Colonel Kimball in his report, April 18, must be incorrect, as General Banks reports on April 17 the capture as already made. *O.W.R.*, series I, volume XV, p. 361.

27. *Ibid.*, pp. 297, 361, 382. It would appear that the confederates intended up to the last minute to offer resistance, for Latham relates how all the live oaks were cut down by the officer charged with the defense in order that they might not intercept the fire from the rifle pits which he had dug on the slope of the hill. Other officers held that the rifle pits

were not so good a cover as the belt of live oaks. In any case the trees were needlessly sacrificed. *Black and White,* pp. 180, 184-85.

28. *O.W.R.,* series I, volume IV, p. 541.

29. *Ibid.,* p. 543.

30. The Kentucky village on the Big Sandy River, usually known as Pikeville, is undoubtedly alluded to.

31. *Ibid.,* volume VII, pp. 883-84.

32. *Ibid.,* volume XII, part 3, p. 901.

33. He sent such warnings on January 25, and 29, 1863, based on reports from various sources. There is corroborative evidence from the union side that raids on the salt works were contemplated at that time, as it was understood that there were only 600 men at Abingdon to guard the works. *Ibid.,* volume XX, part 2, p. 310. For General Jones's warnings see *Ibid.,* volume XXI, p. 1112, and XXV, part 2, p. 599.

34. *Ibid.,* volume XXV, part 2, pp. 599-600.

35. *Ibid.,* p. 600. It should be noted, however, that when General Marshall, who was in command of the Department of East Tennessee, made a raid into Kentucky in April, 1863, General Jones was given due notice that he should keep in special view the defense of the salt works. *Ibid.,* p. 698. They were made his especial charge to be defended in all contingencies. *Ibid.,* p. 797.

36. See his letter of February 1, 1863, to General Floyd, *Ibid.,* p. 603.

37. *Ibid.,* pp. 731-32, 734, 735.

38. The writer had the story from a direct descendant, though it has been printed in the *Bluefield Summit News* (Bluefield, West Virginia), December 21, 1929.

39. *O.W.R.,* series I, volume XXVII, part 3, p. 889; volume XXIII, part 2, pp. 874-75. Jones expressed this view with emphasis to the secretary of war on June 14, 1863, and submitted drawings of such works, drawn by General Buckner.

40. *Ibid.,* volume XXVII, part 3, p. 1056; volume XXIII, part 2, p. 946.

41. *Ibid.,* volume XXVII, part 3, p. 1019. This same general view was expressed by General Buckner, *Ibid.,* volume XXXII, part 3, p. 803.

42. *Ibid.,* volume XXX, part 4, p. 617; volume XXIX, part 2, p. 702.

43. For exchanges of telegrams see *Ibid.,* pp. 699-700, 707, 718, 733, 735, 738, and 756; volume XXX, part 4, pp. 645 and 676.

44. *Ibid.,* p. 753. A rumor of which even President Lincoln took note may be repeated for what it may be worth. On October 23 an Irish refugee reported to a union officer that a few days previously, evidently about October 15 or 16, Governor Smith had addressed the citizens of Richmond, urging the home companies to go to the protection of the salt works. President Lincoln thought that this could only refer to the works near Abingdon. The union report of the incident adds with malicious satisfaction that very few went. *Ibid.,* volume XXIX, part 2, pp. 371 and 376.

45. The salt works had, apparently, been in General Grant's thought for some time, for early the preceding December, while directing him to pursue General Longstreet, he had written to the same general, Foster, "If your troops can get as far as Saltville and destroy the works there, it will be an immense loss to the enemy." *Ibid.,* volume XXXI, part 3, p. 345.

46. *Ibid.*, volume XXXII, part 2, p. 209.

47. *Ibid.*, p. 402.

48. *Ibid.*, p. 472. This response was dated February 26, 1864.

49. *Ibid.*, volume XII, part 3, pp. 287-88.

50. *Ibid.*, volume XXXII, part 2, pp. 450-51.

51. *Ibid.*, part 3, p. 225.

52. *Ibid.*, pp. 596 and 599.

53. A case in point occurred April 28, 1864, as seen in a message from General Buckner to General B. R. Johnson: "If the enemy are still in retreat, send Morgan's dismounted men back to Saltville on the same train on which they went down this morning. Information renders it advisable to cover the salt-works." *Ibid.*, p. 836. Note also the care to leave pickets. *Ibid.*, p. 832.

54. Witness General Burnside's remarks to General Grant on November 4, 1863: "If you think I am holding too many troops in the eastern part of the state, I can easily withdraw them and hold the position with a smaller force; but I am satisfied that presence of the force in this section holds a very large number of the enemy in front of the salt-works, which would be relieved for a movement in this direction or in Virginia to reinforce Lee were we to weaken our force there." *Ibid.*, volume XXXI, part 3, p. 45.

55. To deflect the threatened support from Kentucky under the command of Burbridge, General Morgan took the offensive. Though a failure so far as its direct objective was concerned, the manoeuvre delayed the apprehended incursion into southwestern Virginia for some months. Johnson and Buel, (Edts.) *Battles and Leaders of the Civil War*, volume IV, pp. 423-24.

56. For the account of the fight see *Ibid.*, pp. 478-79. The effort was defeated by a reserve force consisting principally of the county militia. See letter of J. M. Gibson to Governor Vance under date of October 12, 1864, Z. B. Vance Papers, volume V. North Carolina Historical Commission.

For the account of the fight see Johnson and Buel, (Edts.) *op. cit.*, volume IV, 478-79.

57. The various detachments under direction of General Breckenridge had been for some time in Tazewell County, waiting to open up communication with the cavalry of General Forrest through middle Tennessee. Breckenridge had hastily recalled three commands, which, with such home guards as were available, constituted the garrison of the defense of Saltville, *O.W.R.*, series I, volume XLV, part I, pp. 811-12.

58. *Ibid.*

59. *Ibid.*, p. 813. The kettles were about an inch thick at the edge and from two to three inches thick in the bottom so that they were exceedingly difficult to destroy. *Ibid.*, p. 823. General Stoneman found the diameter of the copper tubes in the wells to be the same as that of a twelve-pounder gun, and so, as the most effectual way of destroying them, they were filled with twelve-pounder shells and railroad iron. *Ibid.* See also for the conflict from a local point of view, Summers, *History of Southwest Virginia*, pp. 534-42.

60. *O.W.R.*, series I, volume XLV, part I, p. 817. The engineers were

of the opinion that it would be cheaper and more expedient to bore new wells than to clear out the old ones. *Ibid.,* pp. 813-14.

61. *Ibid.,* p. 841. It is a commentary on war that General Stoneman resented the statement in southern papers that his work of destruction had not been complete, and that 128 good kettles remained after only 788 were broken! He indignantly insists that a thorough destruction had been effected. *Ibid.*, pp. 813-14.

The above account of this raid is based on Johnson and Buel (Edts.), *Battles and Leaders of the Civil War,* volume IV, p. 479 and on the *Official War Records,* series I, volume XLV, part I, pp. 809-17, 841.

62. *Ibid.,* volume XVI, part I, p. 861.

63. *Ibid.,* volume XX, part I, p. 11.

64. *Ibid.,* volume XV, p. 344.

65. *Ibid.,* volume XXIV, part I, p. 338.

66. *Ibid.,* volume XLI, part I, p. 880.

67. *Ibid.,* volume XLI, part 3, p. 936; volume IV, p. 202; volume XVI, part I, p. 973. It is conspicuous that the value of the supplies captured by General Jackson at Winchester and Martinsburg in May, 1862, were carefully estimated. Among them occurs the notation of 350 bushels of salt as worth five dollars a bushel; therefore, the salt spoils were reckoned as worth $1,750. *Ibid.,* volume XII, part I, p. 720.

68. *Ibid.,* volume XLVII, part 2, pp. 476, 485. It is pathetic to read the note addressed by three judges of the inferior court to General McCook at Thomasville, Georgia, just after the surrender, "There is also a small lot of salt here, that is exceedingly scarce with us, and would be of no value to you, which we would also ask for to distribute to the needy." *Ibid.,* volume XLIX, part 2, p. 747. The request came as the result of a public meeting to devise some plan to feed the destitute.

## CHAPTER XIII

1. One of the reports of Commissioner McGehee in Alabama is quite overpowering; unfolded it is literally yards wide, consisting of long sheets painstakingly pasted together (ten feet by two in size), so that one wonders just what principle of bookkeeping underlay such a system and utters a fervent prayer of gratitude that the modern system of typewritten reports protects the future historian from any such records.

2. The documents which are extant on the financial side for Virginia are remarkably brief and simple, but give all the essential figures.

3. As is true for all phases of this subject, the records are remarkably full and satisfying for North Carolina.

4. The writer is adhering to her earlier calculation of the average size of the confederate army, based on figures given by Livermore, *Numbers and Losses during the Civil War in America.*

5. See above, p. 14.

6. "Virginia Documents Relative to the Subject of Salt," Document No. 43, p. 172.

7. It made, as will be recalled, 400 bushels a day at West Bay, (*American Annual Cyclopedia,* 1864, p. 378), which alone could have half supplied the soldiers' needs if the works had been allowed to proceed

uninterrupted, which the union navy did not, of course, allow.

8. *Official War Records,* series I, volume XXIV, part I, pp. 598-99.

9. To estimate only the amount produced in a given state fails to consider the non-residents of that state who were producing the article in that state. The effort is therefore directed to approximate to the amount made for the state.

10. These figures are based on Hilgard's report made shortly after the war in 1873 and may, therefore, be held fairly authoritative, *Final Report of a Geological Reconnoissance of the State of Louisiana,* published in *Geological and Agricultural Resources of Louisiana,* pp. 28-29; and on Veatch, "The Salines of North Louisiana," pp. 73, 59, 67.

11. *On the Rock Salt Deposit of Petit Anse,* (Report of the American Bureau of Mines), p. 18.

12. It must be pointed out that it is difficult to be certain whether Governor Milton's calculation is the usual seasonal allotment or is to provide a minimum for the unusual stringency. Presumably, it does not indicate the usual amount required in this state for the packing season. See Governor's Annual Message, November 17, 1862, with accompanying documents. *Florida Senate Journal,* p. 50.

13. Governor Milton put the fact trenchantly. "Florida manufactures thousands of bushels where other states make hundreds." *Ibid.,* p. 51.

14. The writer has attempted to calculate figures for the production in Florida from the *Official War Records,* endeavoring to combine the production at various points along the coast where there is evidence of salt-making. The utmost that can be claimed, however, is that at least that amount of salt was made. The table appears as follows:

| | | |
|---|---|---|
| District near Tallahassee from the Suwanee to the Choctawhatchee Rivers | 2,000 bushels a day, | *O.W.R.,* series I, vol. XII, p. 703 |
| Lake Ocala | 150 bushels a day, | *Annual Cyclopedia,* 1864, p. 378. |
| West Bay (C.S.A.) | 400 bushels a day, | *Ibid.* |
| St. Mark's Bay | 2,500 bushels a day, | *O.N.R.,* series I, vol. 17, p. 648. |
| St. Joseph's Bay | 200 bushels a day. | |
| St. Andrew's Bay | 1,447 bushels a day. | |
| Cedar Keys | 150 bushels a day. | |
| Kent's Works | 130 bushels a day. | |
| California Inlet | 600 bushels a day, | (1,000 workers). |
| Total | 7,577 bushels a day. | |

15. See above, p. 46.

16. *O.W.R.,* series IV, volume III, p. 250.

17. The valuable report of Commissioner Philips, on which the above statistics are based, is preserved in the report of Z. A. Philips to January 1, 1865, series G, No. 201, Mississippi Archives.

18. *Ibid.*

19. Census of 1860, volume on *Population of the United States in 1860,* p. IV; Livermore, *op. cit.,* p. 21. Of course, the number of soldiers to be deducted is not strictly accurate as Livermore's calculation is for the number of men between 18 and 45 years of age subject to conscription, which full number was not in the ranks at any one time. The number is, however, thought fair for a rough estimate of the number of those whose

salt needs were cared for by the Confederate States government.

20. Alabama Archives.

21. This is the estimate given by Fleming, *Civil War and Reconstruction in Alabama,* p. 158. It is obviously only a rough estimate, for he gives no authorities for his figures.

Fleming estimates 600 bushels a day, but the average through the war was clearly lower than that. The *Charleston Daily Courier* rejoices over 100 bushels a day from the Alabama works on June 25, 1863; Philips, Commissioner for Mississippi, estimates as late as May 22, 1864, that Alabama was then supplying only 600 bushels a week to the state, basing that statement, however, on the opinion of a boat captain. Letter of Philips to Governor Clark, Pettus Papers, series E, volume 65, Mississippi Archives.

22. Fleming, *op. cit.,* p. 160. It must be recorded that these works were destroyed in 1864.

23. It is possible that if one were to make a careful study of the auditor's vouchers from 1861 to 1865, which have been preserved in manuscript form but never printed, an accurate statement of the total amount manufactured for Alabama could be secured, but only a person trained in the keeping of state accounts could handle them satisfactorily.

24. *Charleston Daily Courier,* August 16, 1862.

25. Governor Vance to General G. W. Smith, January 24, 1863, *O.W.R.,* series I, volume XVIII, p. 856.

26. Report of Daniel Worth to Governor Vance, May 6, 1864, *North Carolina Legislative Documents,* 1863-64, Document No. 10, p. 4; *The Correspondence of Jonathan Worth,* volume I, p. 215, gives the total up to December 16, 1862.

27. Woodfin to Ex-Governor Swain, November 17, 1862, Walter Swain Manuscripts, volume IV, North Carolina Historical Commission, and *Legislative Documents,* No. 13, p. 4. He talked of 11,000 bushels a month but that amount was not attained at that time.

28. Woodfin to Governor Vance, July 1, 1863, Vance Letter Book, 1862-63, pp. 343-45. The writer is aware, of course, that she is being very conservative, but careful study has convinced her that she cannot accept the large daily averages given by Worth and Governor Vance for long period calculations, as the work was too often interrupted.

29. *Senate Journal of Virginia,* 1864-65, Appendix to the Governor's Messages, Document No. 8, p. 3. The original is in the Huntingdon Library, but a photostatic copy is available in the Virginia Library. The report was submitted by the Board of Supervisors on January 19, 1865. A report giving exactly the same figures, though ostensibly for a briefer period, from the commencement of state activity to December 7, 1864, is printed in *Virginia Documents,* No. 43, "Relative to the Subject of Salt," pp. 169-171.

30. The following items, it is conceived, are worth reprinting:

| | |
|---|---|
| Exchanged or bartered ....................(also 63,967 given) | 63,957.06 |
| R. N. Clarkson and Company (salt borrowed) .................... | 9,699.03 |
| Negro hire .................... | 4,361.25 |
| Cash .................... | 160.37 |
| Lunatic Asylum .................... | 105.00 |
| Sundry Individuals .................... | 208 |

*Ibid.* p. 171.

31. Census of 1860, volume on *Population of the United States in 1860,* p. IV. The writer is deducting from the total population 1,596,318, the number of soldiers Livermore records as subject to conscription from this state, 116,869, *op. cit.* p. 21. The reader is again warned that this is only a rough calculation.

32. A report of the Joint Committee on Salt to the extra session of the assembly of 1863, indicated a higher daily average of production, 1,613 bushels, for a more restricted period, *Senate Journal and Documents of Virginia,* Extra Session, 1863, Document No. 10, p. 9. At a later period it reached 1,762 bushels a day. An estimate made by Superintendent Clarkson for the period from June 8, 1863, to January 1, 1864, shows the furnaces leased by Virginia as producing 1,400 bushels a day. *Documents of the General Assembly of Virginia,* Document No. 27.

33. This figure is given by Schwab, *The Confederate States of America,* p. 268. The tendency to war exaggeration is shown in a statement that the Virginia wells produced 10,000 bushels a day. *O.W.R.*, series I, volume XXIII, part 2, p. 906.

34. *Virginia Senate Journal and Documents,* Session of 1863-64, Document No. 19, p. 9.

35. The author has calculated these totals from a manuscript entitled, Salt Returns, rendered by Jesse M. Butt, agent, and dated Atlanta, January, 1863-April, 1864, in the Georgia Archives.

36. The figures of the total population in Georgia and the number of males of conscript age are based as before on the Census return of 1860 and on Livermore.

37. Report of Joint Committee on Salt Supply, April 13, 1863, *Senate Journal of Georgia,* Called Session, 1863, p. 127. A Virginia source reported a much smaller amount of daily production for Georgia.

38. The statement of the amount of the yield from the springs in Texas is based on Buckley, *First Annual Report of the Geological and Agricultural Survey of Texas,* prepared in 1874 shortly after the war and so held the most authoritative of the accounts concerning the product of these salines, p. 127.

39. Unpublished manuscripts in the War Department.

40. Computed from the various sums given in the state laws. Though already noted in other connections, they are here again assembled.

Alabama

$150,000, First Regular Session, 1861, No. 29, section 7.

$100,000, Second Regular Session, 1862, No. 38, section 5.

Arkansas

$300,000, for salt, iron, cotton cards, Fourteenth Session, 1862, November 29, 1862, section 1.

$200,000 for salt, Called Session, September-October, 1864, out of the $1,200,000 relief fund.

Georgia

$50,000 to be advanced for encouragement of manufacture of salt, Regular Session, 1861, December 16, 1861, No. 1, section 1.

$500,000 Regular Session, 1862, No. 2, section 1.

.................... Freight on salt to counties from sums not otherwise appropriated, amount not designated. *Ibid.*, No. 46, section 1.

Mississippi

$500,000 out of military fund, Called Session, 1862-63, Chapter XIII, section 1.

North Carolina

$100,000 Ordinances of the Convention, December, 1861.

$35,000 for Saltville works, Session of 1862-63, Chapter 22, section 1.

$150,000 to pay for 50,000 bushels of salt bought of Stuart, Buchanan Company, Regular Session, 1864-65, Chapter 29, section 4.

$200,000 to buy engine and train for transportation of salt from Saltville, *Ibid.*, section 5.

$100,000 for salt commissioner at Wilmington. *Ibid.* (It is not perfectly clear that this is a new appropriation, but the writer reached the conclusion that it was.)

South Carolina

$10,000 to help two stock companies by subscribing stock. Session of 1861, December 21, 1861, No. 4598.

$20,000 for erection of the two magazines, Session of February-April, 1863, No. 4621.

Texas

$50,000 put at disposal of Military Board for manufacture of salt. Session of Nov.-Dec., 1863, Joint Resolution, Chapter VIII.

Virginia

$1,000,000, Adjourned Session, March, 1863, Chapter 17, section 15.

$2,000,000, Session of 1863-64, Law of March 8, 1864, chapter 6, section 9.

Total ........................................................................................ $5,465,000.

41. Governor Brown's letter to Whitaker, commissary general, of August 26, 1862, Letter Books-Governors', 1861-65, p. 337, Georgia Archives; and Brown's order to Colonel B. May, treasurer of the Western and Atlantic Railroad Company, to honor such drafts under the promise that the money would be repaid as soon as the salt was sold. Executive Department—Minutes, 1860-65, p. 333, Georgia Archives.

42. If anything, the governor was niggardly with Mr. Wikle, the agent in Virginia, for the former directed only $5,000 to be sent to Saltville, while further needs were apparently to be met from the revenue derived from the sale of salt. See Governor Brown's letter to Whitaker, dated August 26, 1862, Letter Books-Governors', 1861-65, p. 337, Georgia Archives.

43. Letter of Brown to Whitaker, September 26, 1864, *The Confederate Records of the State of Georgia,* volume II, pp. 728-32.

44. The deduction that this latter sum was repayment on the loan rather than freight seems justified because the state road was carrying private salt gratis. Salt Account Book, Adjutant General's Office, pp. 1-4, Georgia Archives.

45. Although the total is given for the period of September, 1862, to March, 1863, the items also appear, so that it is simple to select totals for the months desired.

46. These figures are as follows:

| | |
|---|---|
| Received for sale of salt for 1862 (Sept.-Dec.) | $55,606.00 |
| Received for sale of salt for 1863 | 233,060.00 |
| Received for sale of salt for 1864 | 246,228.59 |
| Received for sale of salt for 1865 | 73,116.00 |

Salt Account Book, Georgia Archives.

47. The treasurer's yearly reports, where this detail might conceivably have been summarized, seem not to have been published during the war, probably a war economy.

One tabular statement, at least, might be reproduced from the manuscripts:

| | | | |
|---|---|---|---|
| Passage of 4 negroes | $18.00 | Expense on cows | $6.00 |
| Wharfage | 7.25 | Sage and pepper | 4.25 |
| Meat | 11.25 | For graves | 7.00 |
| Extra Work Sunday | 13.25 | Green corn | 16.00 |
| Boxes and bags | 18.00 | Peas | 31.00 |
| Potatoes | 43.25 | Leather | 3.65 |
| Butter | 13.40 | Passage of negroes home | 35.50 |
| Extra wood | 38.50 | Stamps | 20.00 |
| Hauling on Sunday | 16.50 | | |

Abstract of Money Expended by A. G. McGehee, April 1-October 1, 1863, Alabama Archives. An item of nine head of beef cattle on foot costing $1,280 appears in an abstract of October 1, 1863, to January 20, 1864.

The state when it embarked on manufacture was naturally obliged to provide the sacks for the salt. Shorter wrote McGehee on September 12, 1862: "We have a large number of three-bushel sacks, already prepared and many more making, and will have a supply at Saltville, as soon as needed." Executive Correspondence, Alabama Archives.

48. Auditor's Records, Alabama Archives.

49. *Ibid.*

50. Report of D. C. Green, Alabama Archives.

51. Abstract of Benjamin Woolsey, Salt Commissioner, Alabama Archives.

52. Series G, No. 201, Mississippi Archives.

53. *Ibid.*

54. Pettus Papers, volume 58, Mississippi Archives. Freight amounted to $833.75, drayage to $112.50.

55. Pattison's receipts on torn bits of paper, misspelled, in pencil, and in poor French indicate the haste with which he made his purchases in Louisiana. One such item on September 9, 1862, reads as follows: "9 br. post de 56 barils de 50— $28." Series G, No. 201, Mississippi Archives. Pattison was paid $200 for his expenses to Louisiana.

56. Series G, No. 201, Mississippi Archives.

57. *Ibid.*

58. *Ibid.*

59. Various Papers Relating to the Operation of the Salt Department of Mississippi, 1862-65, Mississippi Archives.

A list of the articles which Philips sold after the close of the war shows the vast variety of property acquired by the state in this salt enterprise:

| | |
|---|---|
| 7 mules | 3 draining knives |
| 4 wagons | 1 compass |
| 4 salt-harnesses | 3 chisels |
| 12 traces | 1 bellows |
| 6 harness and 6 collars | 6 blink tooks |
| 1 boiler and 1 engine | 1 balance |
| 1 grist mill | 2 pairs of tongs |

Series G, No. 201, Mississippi Archives.

60. Rowland, *Mississippi*, volume II, p. 592.

61. *Mississippi Senate Journal*, 1862, Appendix, pp. 191-92, 193, 210-11; series G, No. 201, Mississippi Archives.

62. *The Correspondence of Jonathan Worth*, volume I, p. 215. A few weeks later Worth informed Governor Vance that the salt makers, at a meeting requested by the governor, had decided that it cost eight dollars a bushel to make salt at that time, Worth denied any advantage over the others except in the labor of conscripts, which reduced his cost not more than fifty cents a bushel. The letter is dated January 12, 1863. Executive Papers, Z. B. Vance, North Carolina Historical Commission.

63. See Worth's report of May 6, 1864, *North Carolina Legislative Documents*, 1863-64, Document No. 10, p. 5.

64. See Worth's report of January 12, 1863, Executive Papers, Z. B. Vance, North Carolina Historical Commission.

65. North Carolina had a four-fifths interest in the steamer. Vance Letter Books, 1863-65, September-March, pp. 185-86.

66. *Ibid.*

67. "Documents Relative to the Subject of Salt," *Virginia Documents;* Document No. 43, p. 138.

68. *Ibid.*

69. *Florida Senate Journal*, Regular Session, 1863, p. 26.

70. These appropriations have been duly noted earlier in this book.

## CHAPTER XIV

1. Z. A. Philips, the general salt agent of Mississippi, certified to an abstract of the articles sold by him "to the best advantage." He stated that the steam grist mill and saw mill had been burned previously to their sale by some unknown person, and that much of the property was in a "damaged condition and realized prices accordingly," Series G, No. 201, Mississippi Archives.

2. *New York Herald*, October 30, 1862.

3. See Woodfin's report of December 15, 1864, Executive Papers, Z. B. Vance, North Carolina Historical Commission.

4. Woodfin tells of some thirty or forty beef cattle driven off by the confederates and slaughtered just after the raid on Saltville of December, 1864.

5. *Ibid., Legislative Documents of North Carolina*, 1863-64, Document No. 10, p. 2.

6. Message of the governor, *Florida Senate Journal,* 1862, p. 29.

7. A table of importations of salt into Virginia is interesting. The two ports of that state, Norfolk and Portsmouth, imported directly from abroad:

| | |
|---|---|
| 3,946,648 lbs. in 1868 | 3,067,287 lbs. in 1871 |
| 80,290 lbs. in 1869 | 3,321,936 lbs. in 1872 |
| 4,145,211 lbs. in 1870 | 6,247,445 lbs. in 1873 |
| 3,276,249 lbs. in 1874 | |

Hotchkiss, *Virginia,* p. 136.

# BIBLIOGRAPHY

# BIBLIOGRAPHY

## PRIMARY SOURCES

### *Manuscript Sources*

#### GENERAL

Civil War Papers, Quarter-Master's Department, May-September, 1863, In the Archives of the War Department, Washington.

Unpublished Official War Records, in the Archives of the War Department, Washington.

#### STATE

*Alabama*

Abstracts of the Salt Commissioner, Alabama Archives.

Auditor's Record, Alabama Archives.

A. B. Moore Letter File, Alabama Archives.

Executive Correspondence, Alabama Archives.

Series W, Alabama Archives.

Minute Book, Court of County Commissioners, Clarke County, War Period.

*Georgia*

Executive Department-Minutes, 1860-65, Georgia Archives.

Letter Books—Governors', 1861-65. Loose Papers, including one Package of Salt Returns, January, 1863—April, 1864, Jesse M. Butt, agent. (About 75 or 100 papers, weekly reports giving amount of salt received by agent and amounts issued to various counties.) Georgia Archives.

Salt Account Books, Georgia Archives.

*Mississippi*

Pettus Papers, Mississippi Archives.

Series G, Mississippi Archives.

Various Papers Relating to the Operation of the Salt Department of Mississippi, 1862-65, Mississippi Archives.

*North Carolina*

Executive Papers, Z. B. Vance, North Carolina Historical Commission.

Legislative Papers, North Carolina Historical Commission.

Letter Book, 1861-62, Governor Clark, North Carolina Historical Com.

Swain Manuscripts, 1770-1869, North Carolina Historical Commission.

Vance Letter Books, North Carolina Historical Commission.

Walter Clark Manuscripts, North Carolina Historical Commission.

Z. B. Vance Papers, North Carolina Historical Commission.

*Texas*

Confederate Military Board Records, State Library Archives, Texas.

*Virginia*

One Package of Papers Relating to Salt, 1860-65, Virginia Archives.

## *Official Documents*

### GENERAL

Freeman, Douglas S. *A Calendar of Confederate State Papers.* Richmond, 1908.

*Journal of the Senate of the Confederate States of America.* (Published in Senate Documents, 58 Congress, 2 Session, No. 234.) Washington, 1904-05.

*The Statutes at Large of the Confederate States of America.* (ed. James M. Matthews.) Richmond, 1862-64.

*Official Records of the Union and Confederate Navies of the War of the Rebellion,* volumes 17 and 19. Washington, 1903-05.

*The War of the Rebellion: A Compilation of the Official Records of the Union and Confederate Armies.* Washington, 1880-1902.

*Population of the United States in 1860* (Compiled from the Eighth Census). Washington, 1864.

"Proceedings of First Confederate Congress," *Southern Historical Society Papers,* volume 45. Richmond, 1925.

(This is based on the *Journal* of the Confederate States officially published by the United States Government and on newspapers, especially the Richmond *Examiner,* but also the Richmond *Enquirer.*)

## *State Official Documents*

*Journals*

*Journals of the House of Representatives of the State of Alabama* for the sessions 1861-65.

*Journals of the Senate of the State of Alabama* for the sessions 1861-65.

*Journal of the House of Representatives of Arkansas,* 1862.

*Journal of the Proceedings of the House of Representatives of the General Assembly of the State of Florida* for the sessions 1861-65.

*Journal of the Proceedings of the Senate of the General Assembly of the State of Florida* for the sessions 1861-65.

*Journal of the House of Representatives of the State of Georgia* for the sessions 1861-65.

*Journal of the Senate of the State of Georgia* for the sessions 1861-65.

*Journal of the House of Representatives of the State of Mississippi* for the sessions 1861-65.

*Journal of the Senate of the State of Mississippi* for the sessions 1861-65.

*Journal of the State Convention of North Carolina,* 1861-62. Raleigh, 1862.

*Journal of the House of Representatives of the State of South Carolina* for the sessions 1861-65.

*Journal of the Senate of the State of South Carolina* for the sessions 1861-65.

*Journal of the Convention of the People of South Carolina,* 1860-62, Columbia, South Carolina, 1862.

*Journal of the House of Delegates of Virginia* for the sessions 1861-64.

*Journal of the Senate of the Commonwealth of Virginia* for 1861-64.

*Ordinances of the States*

*Acts of the General Assembly of Alabama,* 1861-65.

*Acts Passed by the General Assembly of the State of Arkansas,* 1861-65.

*Acts and Resolutions Adopted by the General Assembly of Florida,* 1861-65.

*Acts of the General Assembly of the State of Georgia,* 1861-65.

*Acts of the State of Louisiana,* 1861-62.

*Laws of the State of Mississippi,* 1861-65.

*Ordinances of the Convention of the State of North Carolina,* 1861-62 Raleigh, 1863.

*Public Laws of the State of North Carolina,* 1861-65.

*Private Laws of the State of North Carolina,* Regular Session, 1864-65.

*Acts of the General Assembly of the State of South Carolina,* 1861-65.

*General Laws of the State of Texas.*

*Acts of the General Assembly of the State of Virginia,* 1861-64.

*Ordinances of the Virginia Constitutional Convention,* April-May, 1861.

*State Reports and Documents*

"Orders and Correspondence Relating to the Seizure of Salt at Apalachicola," accompanying the Governor's message at the Session of November-December, 1862, bound with *Georgia Journal and Documents of the Senate,* Extra Session, 1862, Document No. 10.

*The Confederate Records of the State of Georgia,* Atlanta, 1909.

"Report of the Salt Commissioner at Saltville, Virginia, November 27, 1862," *North Carolina Legislative Documents,* 1862-63, Document No. 10.

"Report of the Salt Commissioner at Wilmington, May 6, 1864," *North Carolina Legislative Documents,* 1863-64, Document No. 10.

*Reports and Resolutions of the General Assembly of the State of South Carolina,* Session of 1862.

*Calendar of Virginia State Papers and Other Manuscripts from January 1, 1836, to April 15, 1869* (ed. H. W. Flournoy). Richmond, 1893.

*Virginia State Documents*

The documents of the Virginia assembly have been remarkably preserved and printed. They are voluminous and invaluable but badly confused as to arrangement and numbering. The same document is occasionally printed in the senate documents under one number and then appears in another place under another number; some are printed without any title; and sometimes papers relating to salt are published with other subjects quite foreign to salt. The writer has done her best to cite clearly enough so that the reader can identify the document. Since many documents appear in one volume of Legislative Documents or are printed with the governor's message, they are treated bibliographically as if they were articles.

"Report of the Joint Committee Appointed to Visit the Confederate Authorities in regard to Supply of Salt," December 20, 1861, Document No. 7.

"Report Relative to the Release of Salt held by the Confederate Government," 1861 (Made late in December, 1861). Document No. 39.

"Report of the Special Joint Committee Relative to Supply of Salt," January 21, 1862, Document No. 16. (Gives the contract for the purchase of 400,000 bushels of salt by Virginia.)

"Report relative to The Purchase of Salt," 1861-62, Document No. III. (Embodies the letter of Stuart, Buchanan Company, dated May 5, 1862.)

"Report of the Joint Committee Appointed to Correspond with the Lessees of the Salt Works," 1861-62. Document No. 47. (Includes the minority report.)

Document No. 1 (No title, but appears as appendix to the governor's message to the special session of September, 1862. It is dated August 5, 1862 and gives the proposition of Stuart, Buchanan Company relative to salt, but includes other matters than salt.)

"Report Relative to the Salt Contracts of Stuart, Buchanan Company," September 29, 1862. Document No. 3.

Document No. 6. (This document is without title. It gives the contract of Governor Letcher with Scott, Stuart, and Palmer for 150,000 bushels of salt and the report of S. B. French and L. R. Smoot on their efforts to get salt for the Virginia state line.)

"Substitute of R. A. Goghill for Joint Resolution of the House Relative to Supply of Salt," March 14, 1863. Document No. 26.

Document No. 6, 1862-63. (No title appears; it proclaims the arrangements for the distribution of salt in Virginia.)

"Contract between Joint Committee of Senate and House of Delegates and John N. Clarkson in relation to a Supply of Salt," March 6, 1863. Document No. 37.

"Minority Report of the Joint Committee of Senate and House of Delegates Relative to the Contract for Supply of Salt," 1862-63. Document No. 38.

"Report of Joint Committee on Salt," March 27, 1863. Document No. 31.

"Report of the Board of Public Works, Acting as Supervisors of Salt Works," 1863-64. Document No. 6. (This document accompanied the governor's message of September 7, 1863.)

"Report of the Joint Committee to whom was referred Report of the Supervisors of Salt Works," Document No. 19. (Probably submitted about September, 1863; it recites the rule of priority for shipment of salt from Virginia.)

"Communication from the Governor of Georgia Relative to Supply of Salt," October 6, 1863. Document No. 6.

"Report of the Joint Committee on Salt in Respect to Salt for Georgia," October 30, 1863. Document No. 10. (Contains letter of the president of the Board of Public Works, Holliday, and correspondence relating to impressments.)

"Report of the Joint Committee on Salt to the General Assembly," October 30, 1863. Document No. 9.

"Report of the Joint Committee on Salt Relative to the Letter from the

Governor of Georgia, 1863. Document No. 18. (It states again the rule of priority for shipment of salt.)

"Biennial Report of the Board of Public Works to the General Assembly of Virginia," January 27, 1864. Document No. 15.

"Report of the Joint Committee on Salt to the General Assembly," February 22, 1864. Document No. 16.

"Report of the Joint Committee on Salt Relative to the Contract with Stuart, Buchanan and Company." Document No. 26. (This is a supplemental report to one made earlier in the session, and is dated February 24, 1864, a fact established by the Journal.)

"Minority Report of the Joint Committee on Salt relative to Supplying the People with Salt," 1863-64. Document No. 28.

"Answer of Superintendent J. N. Clarkson Relative to the Condition of the Salt Works," February 24, 1864. Document No. 29.

"Statement made by Superintendent John N. Clarkson relative to the Operations at the Salt Works." Document No. 27. (This gives the report of progress up to January, 1864.)

"Report of the Committee on Salt Supply." (No number or date appears on the document. It exists as a pamphlet in the Virginia State Library and is the report made by a special committee in regard to mining the salt. The *Journal* enables one to place the date of its submission to the Virginia Senate as February 15, 1864.)

"Documents Relative to the Subject of Salt," 1864-65. (This is a collection of materials bearing on the subject of varied but invaluable character. Document No. 43. It is treated as the other documents though the issue used by the writer appeared as a separate book.)

"Report of the Superintendent of the Salt Works relative to the Capture of Saltville," Senate Document, 1864-65. Senate Document No. 28.

## DIARIES, LETTERS, REMINISCENCES

Andrews, Matthew Page. *Women of the South in War Times.* Baltimore, 1920.

Avary, Myrta L. *A Virginia Girl in the Civil War.* New York, 1903.

Booth, George W. *Personal Reminiscences of a Maryland Soldier in the War between the States.* Baltimore, 1898.

von Borcke, Johann Heinrich Heros. *Zwei Jahre im Sattel und am Feinde.* Berlin, 1886.

Bradlee, Francis B. C. *Blockade Running During the Civil War.* Salem (Mass.), 1925.

Clay-Clopton, Virginia (Mrs.). *A Belle of the Fifties.* London, 1905.

*The Correspondence of Jonathan Worth.* Raleigh, 1909.

Davis, Varina. *Jefferson Davis, A Memoir by his Wife.* New York, 1890.

Dodge, David. "Domestic Economy in the Confederacy," *Atlantic Monthly,* volume 58. Boston, 1886.

Egleston, George Cary. *A Rebel's Recollections.* New York, 1875.

Hague, Parthenia A. *A Blockaded Family.* Boston and New York, 1888.

*Home Letters of General Sherman* (ed. by M. A. De Wolfe Howe). New York, 1909.

Hopley, Catherine C. *Life in the South from the Commencement of the War by a Blockaded British Subject.* London, 1863.

Jefferson, Thomas. *Notes on the State of Virginia.* London, 1887.

Jones, John B. *A Rebel War Clerk's Diary at the Confederate States Capital.* Philadelphia, 1866.

Lunt, D. S. *A Woman's War Time Journal.* New York, 1918.

Moore, Frank. *The Rebellion Record.* New York, 1864.

McMorries, Edward Young. *History of the First Regiment Alabama Volunteer Infantry.* Montgomery, 1904.

"New Salt Manufacture of the Confederate States," *De Bow's Review,* New Series, volume 31.

*Our Women in the War* (Reprinted from the *Weekly News and Courier* (Charleston). Charleston, 1885.

de Paris, Comte. *History of the Civil War in America.* Philadelphia, 1875-1888.

Pickett, George E. *The Heart of a Soldier.* New York, 1913.

Pollard, E. A. *The Lost Cause.* New York, 1866.

Porcher, Francis P. *Resources of the Southern Fields and Forests.* Charleston, 1869.

Preston, Thomas L. *Historical Sketches and Reminiscences of an Octogenarian.* Richmond, 1900.

Pumphrey, Stanley. *Memories,* London, c. 1882.

*South Carolina Women in the Confederacy,* (Records collected by a Committee from South Carolina State Division, U.D.C.) Columbia (S. C.), 1907.

Stevenson, R. Randolph. *The Southern Side or Andersonville Prison.* Baltimore, 1876.

Taylor, Richard. *Destruction and Reconstruction.* New York, 1879.

Taylor, Thomas. *Running the Blockade.* London, 1876.

Townsend, George A. *Campaign of a Non-Combatant.* New York, 1866.

Wise, George. *History of the Seventeenth Virginia Infantry, C.S.A.* Baltimore, 1870.

## NEWSPAPERS

*Arkansas Gazette* (Little Rock), 1931.

*Bluefield Summit News* (Bluefield, West Va.), 1929.

*Charleston Daily Courier,* 1861-63.

*The Charleston Mercury,* 1862.

*The Clarke County Democrat* (Grove Hill, Ala.), 1861.

*The Daily Courier* (Natchez, Miss.), 1861.

*The Daily Mississippian* (Jackson, Miss.), 1862.

*Daily Richmond Enquirer,* 1861, 1862.

*Daily Richmond Examiner,* 1861.

*The Daily Press* (Nashville), 1863.

*The Daily Southern Crisis* (Jackson, Miss.), 1863.

*Daily Vicksburg Whig,* 1861-62.

*Hinds County Gazette* (Raymond, Miss.), 1862.

*Louisville Daily Journal,* 1862.

*The New Orleans Daily Crescent,* 1861.

*New York Herald,* 1864.

*The New York Times,* 1864-65.

*The Richmond Dispatch,* 1861.
*Savannah Daily Republican,* June, 1862.
*The South Carolinian* (Columbia, S. C.), 1862.
*Southern Confederacy* (Atlanta), 1862.
*South Western Baptist* (Tuskegee, Ala.), 1861.
*The Weekly Raleigh Register,* 1861.
*Western Democrat* (Charlotte, N. C.), 1861.

## *Primary and Secondary Sources*

### GEOLOGICAL WRITINGS

For purposes of simplicity and convenience, the geological writings have been grouped together without any effort to discriminate between them as to primary and secondary authority.

Barksdale, Jelks. *Possible Salt Deposits in the Vicinity of the Jackson Fault, Alabama.* Geological Survey of Alabama, Circular No. 10. University, Alabama, February, 1929.

Buckley, S. R. *A Preliminary Report of the Geological and Agricultural Survey of Texas.* Austin, 1866.

Dumble, E. T. *Second Annual Report of the Geological Survey of Texas.* Austin, 1891.

Harris, Gilbert D. *Geological Survey of Louisiana Rock Salt, its Origin, Geological Occurrence, and Economic Importance in the State of Louisiana.* Bulletin No. 7. Baton Rouge, 1908.

Hilgard, E. W. "Memoranda Concerning the Geological Survey," Appendix to the *Journal of the Senate of the State of Mississippi.* Called Session, 1862.

Hilgard, E. W. *Supplementary and Final Report of a Geological Reconnoissance of the State of Louisiana,* New Orleans, 1873.

Hilgard, E. W. *The Salines of Louisiana* in *Mineral Resources of the United States,* Geological Survey. Washington, 1883.

Lucas, A. F. "Rock Salt of Louisiana," *Transactions of the American Institute of Mining Engineers,* volume XXIX. New York, 1899.

*On the Rock Salt Deposit off Petit Anse; Louisiana Rock Salt Company.* (Report of the American Bureau of Mines.) New York, 1867.

Owen, David Dale. *Second Report of a Geological Reconnoissance of the Middle and Southern Counties of Arkansas.* Philadelphia, 1860.

Patten, J. H. *The Natural Resources of the United States.* New York, 1894.

Phalen, W. C. *Salt Resources of the United States,* United States Geological Survey, Bulletin No. 669. Washington, 1919.

Phalen, W. C. *Technology of Salt-Making in the United States,* Bureau of Mines, Bulletin No. 146. Washington, 1917.

Phillips, W. B. *The Mineral Resources of Texas,* Bulletin of the University of Texas, No. 29. Austin, 1914.

Switzler, William F. *Report on the Internal Commerce of the United States,* Appendix on Virginia. Printed in *House Executive Documents,* 49 Congress, 2 Session, volume 7, part 2. Washington, 1886.

Thomassy, M. J. Raymond. *Géologie Pratique de la Louisiane.* New Orleans, Paris, 1860.

Thomassy, M. J. Raymond. "Supplement à la Géologie Pratique de la Louisiane," Bulletin *de la Société Géologique de France,* Tome 20, 2e série. Paris, 1863.

Veatch, A. C. *The Salines of North Louisiana, A Report on the Geology of Louisiana,* part VI. Baton Rouge, 1902.

Watson, Thomas L. *Mineral Resources of Virginia.* Lynchburg, 1907.

## GENERAL SECONDARY AUTHORITIES

Allan, William. "Jackson's Valley Campaign," *Southern Historical Society Papers,* volume 43. Richmond, 1920-23.

*The American Annual Cyclopedia,* 1861-65. New York, 1862-65.

Ashe, Samuel A. *History of North Carolina.* Raleigh, 1925.

Ball, T. H. (Rev.) *A Glance at the Great South East, or Clarke County, Alabama, and its Surroundings.* Grove Hill, Alabama, 1882.

*Battles and Leaders of the Civil War* (Edts. Johnson, R. U. and Buel, C. C.). New York, 1887-88.

Bonham, Milledge L. "The British Consuls in the Confederacy," *University of Columbia Studies,* volume 43. New York, 1911.

Brevard, Caroline May. *A History of Florida.* Deland, Florida, 1925.

Cartland, Fernando G. *Southern Heroes or the Friends in War Time.* Cambridge, 1895.

Davis, William Watson. *The Civil War and Reconstruction in Florida.* New York, 1913.

Fleming, Walter. *Civil War and Reconstruction in Alabama.* New York, 1905.

Garner, J. W. "The State Government of Mississippi During the Civil War," *Political Science Quarterly,* volume 16. New York, 1901.

Hamilton, Joseph Grégoire de Roulhac. *Reconstruction in North Carolina.* New York, 1914.

Head, T. L. Salt Works in Clarke County (Unpublished manuscript in the Alabama Archives).

King, Grace. *Jean Baptiste le Moyne Sieur de Bienville.* New York, 1892.

Laidley, W. S. *History of Charleston and Kanawha County, West Virginia and Representative Citizens.* Chicago, 1911.

Latham, Henry. *Black and White.* London, 1867.

Livermore, Thomas L. *Numbers and Losses in the Civil War in America, 1861-65.* Boston and New York, 1900.

Matthews, M. M. A Sketch of Historic Clarke County. (Manuscript in the Alabama Archives.)

Maury, M. F. *Physical Survey of Virginia.* Richmond, 1878.

McMaster, John B. *A History of the People of the United States During the Administration of Lincoln.* New York, 1927.

Miller, L. D. *History of Alabama.* Birmingham, 1901.

Moore, Albert B. *Conscription and Conflict in the Confederacy.* New York, 1924.

Moore, John W. History of North Carolina. Raleigh, 1880.

Owen, Thomas M. *History of Alabama and Dictionary of Alabama Biography.* Chicago, 1921.

Rowland, Dunbar. *Mississippi: Comprising Sketches of Counties, Towns, Events, Institutions, and Persons*. Atlanta, 1907.

Schwab, John C. *The Confederate States of America*. New Haven, 1913.

Shotwell, Walter G. *The Civil War in America*. London, New York, 1923.

*The South in the Building of the Nation*, volume V. Richmond, 1909.

Summers, Lewis P. *History of Southwest Virginia*. Richmond, 1903.

Weeks, Stephen B. *Southern Quakers and Slavery*, The Johns Hopkins University Studies, Extra volume XV. Baltimore, 1896.

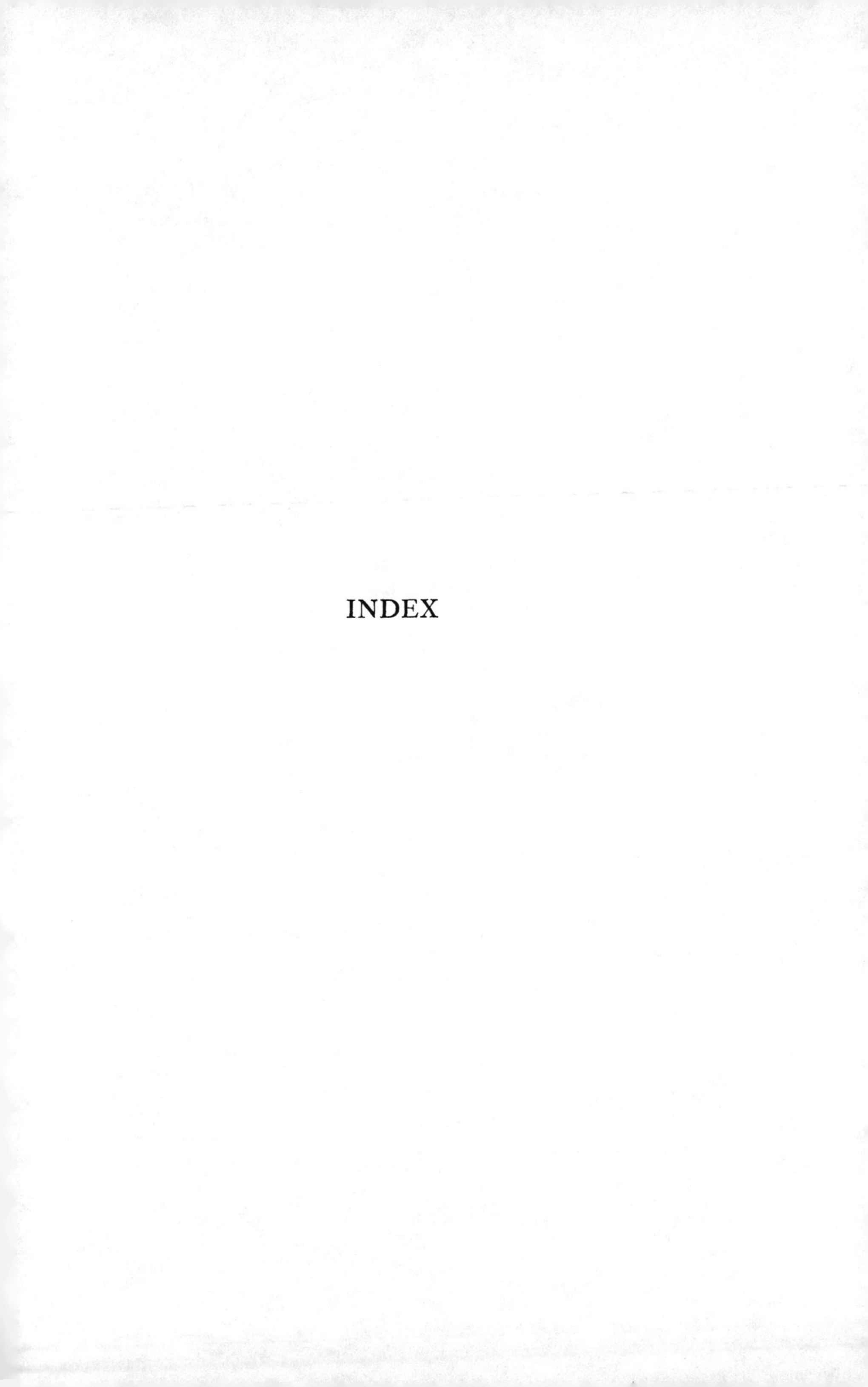

# INDEX

# INDEX